TIM RADFORD is a fr... *Guardian* for thirty-two ... letters editor, arts editor, ... won the Association of Bri... writer of the year four time... ... ... for the International Decade ... ...tural Disaster Reduction. He has lectured about science and the media in dozens of British and foreign cities. He has written one book, *The Crisis of Life on Earth*, and edited two books of science writing for the *Guardian*.

From the reviews of *The Address Book*:

'Radford [is] one of our finest science writers ... Disguised as a survey of popular science, masquerading as a memoir, Radford's book builds into a complex and compelling linguistic, poetic, scientific, spiritual and historical survey'

IAIN FINLAYSON, *The Times*

'Tim Radford is a valuable witness because he is a balanced man, at home in science, respectful, but not intoxicated by it. His beautiful, meditative book is a surprise in these clamourous times: one good deed in a naughty world'

PETER FORBES, *Guardian*

'Cultured and at times quietly amusing, his writing is not only a triumph of plain speaking; it is often beautiful'

CATHARINE MORRIS, *Times Literary Supplement*

'Readably personal slice of popular science ... An early clue that this will be a fun ride comes when Radford moans that post-codes have ruined the romance of addresses ... Radford packs an awful lot of fascinating stuff into a small space'

... ... *Evening Standard*

By the same author

The Crisis of Life on Earth

TIM RADFORD

The Address Book

Our Place in the Scheme of Things

FOURTH ESTATE • London

Fourth Estate
An imprint of HarperCollins*Publishers*
77–85 Fulham Palace Road
Hammersmith
London W6 8JB

This Fourth Estate paperback edition published 2012
1

First published in Great Britain by Fourth Estate in 2011

Copyright © Tim Radford 2011

Tim Radford asserts the moral right to
be identified as the author of this work

A catalogue record for this book is
available from the British Library

ISBN 978-0-00-735629-4

Set in Minion by G&M Designs Limited,
Raunds, Northamptonshire
Printed and bound in Great Britain by
Clays Ltd, St Ives plc

MIX
Paper from
responsible sources
FSC www.fsc.org **FSC**™ **C007454**

FSC™ is a non-profit international organisation established to promote
the responsible management of the world's forests. Products carrying the
FSC label are independently certified to assure consumers that they come
from forests that are managed to meet the social, economic and
ecological needs of present and future generations,
and other controlled sources.

Find out more about HarperCollins and the environment at
www.harpercollins.co.uk/green

To my family

CONTENTS

ONE

The Number and the Street

Whose House is it Anyway?

I miss my little midnight companion, the death watch beetle. Years ago, in the study late at night, in the spring or the autumn, I would sometimes hear him tick, three or four times in a row, like a tiny clock, calling vainly for a potential mate. Right now, it is high summer. I sit in an old study with brick floors, a rough plaster-and-beam ceiling and oak-panelled walls, obscured before me and to my right hand by oak shelves of books, journals, collections of magazine cuttings and bundles of notes; and for the most part I sit in silence. Through a window to my left hand, divided by little diamonds of lead, I can see an unruly rose, some assertive honeysuckle and some leaves of a spiky *Mahonia japonica*, and a few yards beyond these warring shrubs the conical tip of a clipped yew, and beyond that again, the highest branches of a silver birch, and beyond those, only a clear blue sky.

In spring, or in the early morning, I could expect to hear the clamour of a wren, the song of a blackbird or thrush, the chatter of a magpie, the murmuring of wood pigeons and collared doves or the mewing of herring gulls. But right now, this minute, with the late morning sun making patterns on the linen curtain over the window to the left of my desk, the only noise is the muted clatter of the keyboard.

The trees are mine, or rather ours, in a limited sense: for many years we nursed them in pots in the tiny garden of an outer-London home, and then carried them to this house in the snowbound winter of 1985, to plant them at the first thaw and watch them begin their push towards the sky. The room in which I sit and the house and gardens beyond it are mine – or rather ours – in the technical sense that a mortgage has been cleared, and the deeds that indicate our possession rest in the vault of a solicitor. These bits of paper establish this address, and my place in it, and seem to answer a very old and intermittently troubling question: where am I?

At the start of each school term, at the age of about ten, I did something that I suppose a million other ten-year-olds have done: I wrote my name in an exercise book, along with my house number and street. I then added the name of the suburb, and the city. Then, for good measure, I named the administrative region in which my city stood, and just to make sure, the country. And then – where did I imagine I might lose this book, and who would find it? – I wrote 'the Earth', and just in case that wasn't precise enough, I added 'the solar system'. At some point in the performance of this ritual, I decided I had better make absolutely sure, and appended the triumphant conclusion 'the universe'. Later on – much later on – I realised that to be truly pernickety I should have also included the continent in which my country counted itself, the hemisphere I happened to be in, and the galaxy of which the Sun is but one mild little spark, before concluding with the cosmos itself, the whole bag of tricks.

Even at the time, it seemed an obsessive little ritual, but ten-year-olds are embarrassed about neither obsession nor rite. Across the interval of the decades, however, it occurs to me that it might have been my first independent search for answers to questions posed consciously and unconsciously by everyone in every culture, and in every generation: who am I? Where did I come from? Why am I here? Where am I going? To answer questions like that, you have to start somewhere. You start with the question: where am I now? Where I come from is a clue to who I am, and where I find myself is a point on a trajectory through space and time. Place is a powerful part of identity.

I have a precise postal address, but I do not know where I am. I am sitting still, but I am also hurtling eastwards at an estimated six hundred miles – more or less 1,000 kilometres, to use the internationally agreed scientific unit – an hour. On the Equator, I would be going even faster than sound: towards the dawn at a thousand miles an hour. So even when I sit at home, I am a moving target. The house in which I live is built upon solid rock, but sandstone crumbles perceptibly, and the coast upon which I live is eroding. It is also sinking. I am going down in the world. One day, the ruins of this house will ooze into the sea. The ground beneath my feet sinks and heaves and distorts with episodes of drought and flood and frost. I do not see it, but over the years I observe house doors and garden gates that almost jam, or bolts that don't quite fit their hafts, and I conclude that the surface that underpins them has shifted again. So not only do I not know where I am, I do not always know whether I am up or down. The county in which I have made my home is a byword for respectability and order, at the southern edge of a small and densely populated temperate island in which the water runs hot and cold and the trains quite often run on time. But only a few thousand years ago, this county was part of a continent, and a few tens of thousands of years ago it was near the frontier of vast, approaching glaciers, and a few hundreds of thousands of years before that, lions and rhinoceroses roamed its grassy plains.

The climate is not the only thing that has changed: the mainland of Europe is also moving, and Britain is inching away from America at the rate that fingernails grow, and taking my home with it. I move towards the dawn because the planet is rotating on its axis, but it is also doing something else: it is whizzing around the Sun in a huge ellipse, at very roughly thirty kilometres a second. The Sun, too, is moving: it is after all a modest main sequence star – a star like billions of others, and with billions of years of life left in which to burn steadily – near the rim of a galactic disc which is itself spinning on a central axis, so the Sun proceeds around the centre of the galaxy, moving up and down like a horse on a fairground carousel, at a stately two hundred kilometres a second. The galaxy is moving too, but how fast and in which direction depends upon which other galaxy you choose to measure from.

So when I sometimes say I don't know where I am, I am not kidding. I do not have an answer. Nor do I feel alone in asking the question. 'Where am I?' could be the oldest question in history. For the moment, at least, I have an answer of sorts. I am in a house that has a very precise geography, over which I have some temporary control. Desk and chair, the books on the shelves, the rug on the floor, the calendar on the wall, are there because I put them there, and I am at liberty to move them. They are intimate landmarks, places from which, unconsciously, I may take a bearing. I can place myself in relation to the walls, to the rooms, to the floors and ceiling. In this house, I know where I am. There is a 'here' in which I can locate myself. It is a place in which I may choose to stay for hours, or days, without running the risk of being moved on by the police. It is, at the moment of writing, my home. There is a legal document that describes the property, and names me as a joint owner.

But I do not feel as though I own the place. I feel as though I have been given the job of looking after it. That is because I had no part in its design, or its construction, or its extension, or its most recent alteration; nor can I change its exterior or substantially redesign its interior, to make it mine, because the building is listed, Grade II, and the death watch beetle has more right to tenure, and potentially more freedom to amend its structure, than I have. His natural habitat is stained oak and old panelling, and mine is not. If he survives, he deserves his home behind my dictionaries and reference books not just because that is his ecological niche, but also because he withstood the fiercest chemical assault to maintain his place in my study, if not in my heart. The house was drenched with woodworm, furniture beetle and fungal killers when we moved here early in 1985, and the study was attacked again, in a second intense chemical foray, when it became clear that the first assault had left survivors. Ironically, for weeks afterwards, I could not tolerate more than a few seconds' exposure to the fumes that lingered in the room. But *Xestobium rufovillosum* made it through: he clung on in a bunker deep in some timbers where the murderous chemical cocktail could not reach, and for some time afterwards I would occasionally hear him beating his head against the wood. And then finally his tick fell silent.

His links with the house predated mine. His ancestors may have moved in with the first lumps of green heartwood – probably already fungus-enfeebled – when some eighteenth-century artisan began to build a cottage that is now the kitchen, almost three hundred years ago. Or perhaps the death watch beetle migrated, involuntarily, from a church in France or a dismantled folly somewhere in England around 150 years ago, when some ambitious tenant converted an old mill or coachman's cottage into a gentleman's residence, complete with higgledy-piggledy, hand-me-down panelling that varies from room to room, stained glass, beams, tiled and carved fireplaces and distinctive windows of the kind sometimes called Victorian gothic. The study, with its imperfectly fitted wooden panels and uneven plaster, is a kind of untidy rehearsal for the entire house. You step in through the front door into a modest porch. Two doorways open off the porch: one into the study. You have been there already. The other leads immediately into a hall floored with old brick and lit by a huge Victorian window.

From this roomy half-panelled hall, a staircase leads to an assortment of bedrooms, a bathroom, a library and the attic. A winding corridor leads away to the front of the house and a large drawing room and a music room. The curve of this corridor suggests that the original property may have been associated with something round, perhaps a windmill, in the eighteenth century. One thick oak door opens onto a panelled dining room. Another thick door opens into yet another small lobby and a bathroom and then to what would once have been the stable yard. Another passageway winds down a cramped staircase into a large and partly subterranean kitchen that in turn has doorways into a pantry, a larder and a large cellar. I have an auctioneer's poster dating from 1872 which describes the house as a

> very valuable and desirable villa residence in the Parish of St Clement, Hastings, delightfully and romantically situated on the West Hill, in a well sheltered position, about 360 feet above the level of the sea; the house faces west and commands unrivalled and extensive sea and inland views of great beauty.

It doesn't command extensive views any more: you can't see the sea for all the other, more recent houses that have colonised the hill. But the advertisement provides me with a second set of coordinates: I am located in time, as well as space. The property has height, width and length, but it also has duration, and the auctioneer's advertisement and the calendar on my wall supply two points on a temporal continuum. These two points become my starting point for a sedentary exploration, an immediate geography, an imaginary journey, and I quote these details to remind myself how little the place has changed in 140 years.

The music room is a twentieth-century addition, and so is one bathroom. A little conservatory now connects and simultaneously insulates the music room and dining room. Otherwise, the man who put the place up for auction at the Royal Swan Hotel, Hastings, on Saturday, 7 September 1872 would still find himself entirely at home. He would of course marvel at the electric lighting, he would appreciate the plumbing, he would certainly applaud the red, gas-fired Aga that heats all the household water and bakes all its food; the telephone points and broadband links would take some explaining. He would also ask about the corners in the dining room and sitting room where the oak panelling abruptly comes to an end, and I would then have to tell him about our other mutual tenant, *Serpula lacrymans*, or dry rot. This little fungus is another long-term investor with a taste for old houses, and by the time we moved in, it had swelled into a science-fiction monstrosity, had spread its tendrils and had devoured the timbers on either side of an adjoining wall. But the person who owned this house in 1872 would still know the place, and recognise it as his. So, if it is his, in what sense is it mine? Did it shape him, or did he shape it? If he shaped it, then what have I done to it, and what has this house done to me? Can we both, to put it another way, really have the same address?

The answer is, of course not. His habitation was of its time, and so is mine. We all of us occupy more than one home at a time; we all of us have more than one set of geographical coordinates. British listeners and viewers go to Ambridge to keep up with the Archers, or to Albert Square to observe the EastEnders. So much, so obvious, but along with the broadcast soap operas, somewhere in our memories are thousands of other destinations imagined by someone we never met, and never thought much about: the Never-Never, South of the Border, Oz, Middle Earth, the City of Dreadful Night, the Land of Nod, the Little

House on the Prairie, the Golden Road to Samarkand, Camelot, Vanity Fair, Pooh Corner, Ruritania, Shangri-la, Utopia, Barchester, Metropolis, Tuxedo Junction, Penny Lane and Blueberry Hill. Some of them have a tenuous connection with reality, and have become embedded in the global cultural canon, but for all most of us will ever know, they might also be fiction: Ur of the Chaldees and Babylon and Nineveh and the Promised Land Flowing with Milk and Honey; Troy and the walled city of Jericho; Sodom and Gomorrah; Xanadu and El Dorado. They have a ghostly reality in a disorderly index; they make up a directory or a gazetteer of places that may now have no palpable existence, but remain stubbornly addresses in imagination's atlas.

So my forgotten predecessor inhabited the house I now live in, but in 1872 he might also have sat Under the Greenwood Tree, in Wessex, or moved to Erewhon, in the southern hemisphere, because Thomas Hardy and Samuel Butler that year each published a novel that was to become a fixed address in the imagination's gazetteer.

Whoever now lives in Derby Street, Devonport, Auckland, New Zealand – the first lines of the address I would write in my exercise books more than five decades ago – occupies a different place in a different world. That house was, and I believe still is, a detached wooden villa with a corrugated iron roof and a front veranda and the stump of an old pepper tree in the front garden, and it is still instantly recognisable as 'our' house; but whoever lives there now also inhabits a different set of superimposed geographies. I lived there in 1956: I knew about, although never visited, Peyton Place, the bestseller of the year; I could have moved into the Towers of Trebizond, with Rose Macaulay, but I did not. I was fifteen at the time and I had of course, like almost every other teenager in the English-speaking world, just found a new place to dwell: it was down at the end of Lonely Street, in

Heartbreak Hotel. Elvis Presley's song that year reverberated around the planet: it also reverberated within the damp cement changing sheds excavated under the road at Cheltenham Beach, Devonport. This grimy facility – it smelt of dried seaweed, wet sand and stale urine – provided a cold-water shower and toilet for beach users, and a place to hang sandy clothes and wet towels, but more potently, it also provided an impromptu echo chamber in which one particular tiny assembly of exuberant teenagers could improvise their own stumbling versions of what was to become an anthem for gloomed youth.

The emergence of Elvis Presley in 1956, along with James Dean's debut in *East of Eden* and the worldwide introduction to Bill Haley's soundtrack 'Rock Around the Clock', over the credits of the movie *The Blackboard Jungle* in 1955, is now presented by social historians as the moment of an epic shift in the Western cultural continuum – the birth of rock and roll. That's what it seemed like at the time, too. But I cannot hear Presley's version of 'Heartbreak Hotel' without also hearing tuneless voices in a shadowy concrete chamber, a few paces from a stony square of grass ringed by evergreens that opened onto a cement promenade, and then onto a crescent of coarse sand and shell, that fell gently towards a huge expanse – at low tide – of mud and ankle-deep water, before falling away again into a shipping channel, which was bounded a few miles beyond by the seemingly symmetrical outline of the volcanic island of Rangitito, that framed the entrance into Auckland harbour.

We become more or less what we have always been, which might be why we take so little notice of the ways in which we have changed, which in turn might be why, in so many of the chambers of the museum of memory, we find some empty scaffolding, a blank wall, a pail and broom, and a laconic notice that describes the picture that must once have hung there. The unhappy, the abused, the insecure and the fearful are tormented by their memories, and want to forget. Happy people are secure, and securely unaware of their happiness, serenely enjoying each moment, without overtly wishing to remember. The changes, the disjunctions, the Alice in Wonderland moments when we tumble into a new environment, stay with us, but the intimate, physical, moment-by-moment experience of humdrum contentment blurs immediately and fades swiftly. Even so, some memories survive, with a sensual intensity that can be evoked unexpectedly by a smell, a sound, a set of words, a name, a snatch of song, or even a moment in the sun or the rain.

Such recall can happen on some hot, perfect day in the Mediterranean, one of those days when the Sun's rays burn through the fabric of a cotton shirt and begin to caress the shoulder muscles with the force of a physiotherapist's thumbs; because on a summer day in Auckland the Sun could hammer the city hard enough to set up a heat-shimmer on the concrete lanes at the centre of the road, and melt the tar strip on either side. It can happen halfway up Mount Fuji in Japan, at a picnic spot marked by lumps of scoria: fist-sized lumps of rapidly ejected volcanic rock so hot and so gas-filled that when it cooled and fractured it was as full of holes – the technical term is vacuoles – as a sponge, but jagged and unforgiving to the touch; because Auckland, of which Devonport was fifty years ago a poor suburb, was built on dozens of extinct or perhaps just dormant volcanoes, all of them low, conical and piled high with scoriaceous rock. It can happen in summer rain: Auckland was a rainy place – it still is – with twice the precipitation of London, and an enduring memory is of a childhood barefoot in the rain: a warm rain, all too often kept off by an uncomfortably sticky waterproof. It can happen when I stand under a pepper tree in Claremont, California – the Peruvian pepper, *Schinus molle*, a pungent but not particularly common ornamental species that leaves a film of resin on the fingers: my brother and two sisters and I grew up watching the world go by from the gnarled branches of an old pepper tree in the front garden.

My memories of the house at Derby Street, Devonport, are not easily separable from the memories of the little world I knew beyond the house, and this untidiness serves to illustrate the difference between a house and a home: a house is a box in which you shelter, eat, sleep, love, procreate, work and even die, but you could have done the same things in the box next door, or in the next road, or in the next suburb. A home is where you become yourself, and you may leave it, but it does not so easily leave you. The furnishings in Derby Street were – by twenty-first-century standards – modest enough, but there were maps everywhere, because we were the children of a geographer: piles of old *National Geographic* maps; maps of the British Empire with India and Malaya and Canada and Australasia and half of Africa in pink; maps of pre-war and wartime Europe; economic maps intended for schoolroom walls that showed Australia with little pictures of sheep, and ingots of iron, and stooks of wheat; relief maps; maps marked with contours; maps that showed only population density, or rainfall, or vegetation, or prevailing winds; maps of the world's trade routes; and even military maps of Pacific islands printed on silk that subsequently served as scarves.

The furnishings also included bookcases, from which all four of us read indiscriminately, along with whatever we had picked up from the local lending library: on a rainy day one of our parents might enter the living room to find all four of us, cross-legged or lying on our stomachs, following the adventures of Richmal Crompton's William, or Captain W.E. Johns's Biggles, or Captain Hornblower, or discovering What Katy Did and why Anne loved Green Gables, or lost in the ten volumes of Arthur Mee's *Children's Encyclopaedia* or struggling blankly with the Everyman edition of Herodotus, or a novel by Graham Greene.

It was there that we – all of us – discovered that the narrow pages of a book offered a doorway into the wider world, and through these doorways we began to explore and colonise those other places, those other domiciles, fortresses, redoubts, camps and caravanserais that make up an identity's address book.

The glib phrase for such places is 'in the childish imagination', but those landscapes we discover in books are as real to us as any we might happen to see as tourists or townsmen later in life. We do not 'see' reality, we construct it in our heads a fraction of a second after a pattern of electromagnetic wavelengths has hit the retina and set up traffic in the optic nerve. We make a picture of a cliff-face, a city or a log cabin from the reflected radiation perceived by our eyes; but this is also what we do when we read words in a book. When I first heard about the great, grey, greasy banks of the Limpopo, all set about with fever trees, I had no idea at the time that the Limpopo was a real river, but it stayed with me as a potent location all the same, for me and for the Elephant's Child in the *Just-So Stories* by Rudyard Kipling. When, years later, I pursued Kipling through the cities of imperial India, I had no idea of what a newspaper office might be like, but I picked up a very clear idea of one from 'The Man Who Would be King'. By the time I read – and loved – Kipling's extraordinary short stories based in Sussex, I knew that Sussex was a county or administrative region of southern England, and I even knew the words of the march sung by the Royal Sussex Regiment during the First World War, but it never occurred to me that I might live there. These locations of the imagination that grow from the printed pages of books also included addresses that lodged for decades, and perhaps for a lifetime, in the mind: Rat's riverside home in *The Wind in the Willows*, along with Mole End, 'where a garden seat stood on one side of the door, and on the other, a roller', and above all the home occupied in the Wild Wood by Badger, where 'the ruddy brick floor smiled up at the smoky ceiling; the oaken settles, shiny with long wear, exchanged cheerful glances with each other; plates on the dresser grinned at pots on the shelf, and the merry firelight flickered and played over everything without distinction'.

It was from Kenneth Grahame's extraordinary book (extraordinary because he was the Secretary of the Bank of England, and he wrote lines that became quoted throughout the English-speaking world, along with a chapter that provided a record title for Pink Floyd) that I learned that 'the South' was a magic place of romance and warmth. 'What seas lay beyond, green and leaping and crested! What sun-bathed coasts, along which the white villas glittered against the olive woods! What quiet harbours thronged with gallant shipping bound for the purple islands of wine and spice, islands set low in languorous waters,' Rat reflected, in the chapter 'Wayfarers All'. It says a lot for the power of literature – it says a lot for Kenneth Grahame – that Provence and the Côte d'Azur and the Ligurian coast came alive in such sentences, and exerted their collective magic, and made me identify, at ten years old, the south as a place of warmth, and aroma, and bliss. It also says a lot for the power of the imagination that it didn't worry me in the least that, for a New Zealander, the south was the direction of increasing cold, wind, rain and finally frost and snow, and that the Sun occupied the northern half of the sky. And I don't think I noticed at the time that I already lived on a sun-bathed coast lined with white villas glittering against the evergreen wood, and that Auckland was screened by purple islands, set low in languorous waters.

Perhaps it was the absence of wine and spice; perhaps it was the oblivion of familiarity; perhaps it was the force of the printed word – but I never noticed quite how Mediterranean my own home suburb was until I had left it for a couple of decades. More probably, it was because all Antipodeans of European origin, at that date, enjoyed an imported culture.

Our films came from Ealing, Pinewood and Hollywood. Most of our weekly and monthly magazines arrived, months late, by sea from the United Kingdom. Our household had piles of the *National Geographic* magazine, others the *Saturday Evening Post*. We read the books that any Londoner might know. I learned about Squeers's cruel Dotheboys Hall in Yorkshire, by taking *Nicholas Nickleby* with me on the open deck of a wooden ferry-boat to school in the mornings, looking up occasionally when porpoises surfaced to accompany the commuters across the harbour, or when a converted Sunderland or a Solent flying boat carrying passengers from Sydney or Fiji touched down at the waterfront air terminal half a mile distant, the keel of its beautiful, curved hull suddenly creaming the smooth blue water and sending up a fine white spray before it slowed, slumped and taxied to the mooring. These are things I may never see again, and if they were before me now I would snap my book shut and read the landscape's story as greedily as I then read *Moby-Dick* or *Bleak House*. I probably thought my surroundings were of no great romantic consequence, and I preferred instead to be with Ishmael and Queequeg in New Bedford, with its streets full of 'Feegeeans, Tongatabooans, Erromanggoans, Panangians and Brighggians'. This now seems to have been perverse of me: I went to school with, and played rugby with, children from Fiji, Tonga, the Cook Islands and Eastern Samoa, and thought nothing of it. I inhaled the sweet, salty air of one of the world's least polluted cities and longed instead for Dickens's London with its smoke 'lowering down from chimney-pots, making a soft black drizzle, with flakes of soot in it as big as full-grown snow flakes'. I had, of course, never seen a snowflake.

In childhood, one tends to think that one's surroundings can be looked at later, when there is nothing else to read. The understanding that landscape, too, is a thing you must read and understand, because it changes with time, comes later. Trees are felled, scrublands cleared, marshes are drained, wooden houses are torn down and replaced by brick boxes, green fields become car parks. Naturally, most childhood memories are episodic and confined: confined to a street and the interiors of one or two houses; confined to a regular route to school or to a friend's home. My horizons were somewhat wider. I had a bicycle, and used it to deliver morning newspapers from the age of twelve, until I left school to begin work in a newspaper office; I also used the same machine to deliver telegrams after school, at weekends and school holidays, and in consequence got to know every street in Devonport. Some had Maori names. Some were the names of Victorian and early-twentieth-century England: Victoria Road, Albert Road and Allenby Road, Abbotsford Terrace and Jubilee Avenue. Some were simply descriptive: Domain Road led to the domain, or public recreation ground, Lake Road carried traffic north to Lake Pupuke in Takapuna. One bore the name of a Greek muse, Calliope, and it led to the Devonport Naval Base.

Only now, on looking at a street map of Devonport, have I noticed two streets that I must once have known, but now do not remember: Hastings Parade and St Leonards Road. Hastings is a common enough name in the English-speaking world. There are towns named Hastings in Michigan and Nebraska in the United States; and in Ontario, Canada; there is even a Hastings in New Zealand, the location of that country's most damaging recorded earthquake. There are streets, avenues and roads called Hastings all over Britain, Australia, New Zealand and Canada, many of them certainly named after Warren Hastings, the eighteenth-century adventurer caught up in the colonisation of India, but the street in Devonport was certainly named after Hastings, East Sussex, because the next street was called St Leonards, after the adjoining town in East Sussex. So there is a tenuously prophetic connection between the place in which I grew up, and the place in which I live now. The only impact of the discovery is to remind me once again of the intense physical difference between the gloriously warm, sparkling, almost landlocked harbour of Auckland and the muddy green chill of the English Channel. Memories mislead: Auckland must have been overcast and rainy for weeks on end. And I know from almost daily observation over decades that the English Channel, too, sparkles in the sunlight. But we keep a template of reality in our heads, and when we aren't looking we remember a place as a snapshot taken at some contented moment.

Memories fade, telescope or mutate, but I find on my occasional returns to Devonport that nothing seems to have changed at all: everything is where it always was, only a lot more expensive, and each time I return I can find my way, without any pause for recall, to Derby Street. It wasn't where I was born, but it provides me with a starting point: all shelters before that address were just that, places where we stayed. But in 1947 we moved into a weatherboard villa at latitude 36 degrees 49 minutes and a bit South, 174 degrees 48 minutes and something East, and stayed there until I was seventeen. Neither longitude nor latitude is anything more than a convenient way of pinpointing an address on the globe. Longitude 174 sweeps north and south across empty Pacific Ocean, crossing only a shallow reach of Antarctica and eastern Siberia before it touches the poles. The line of latitude, too, spans mostly sea: it slices across the Pacific, Atlantic and Indian oceans, cutting through Chile just north of Concepción, and emerging from Argentina south of Montevideo. It misses South Africa altogether, and bisects Australia only in Victoria, a little north of Melbourne. It is an address distinguished only by its loneliness. Kipling's salute to Auckland in 1897 says it all:

Last, loneliest, loveliest, exquisite, apart –
On us, on us, the unswerving season smiles,
Who wonder 'mid our fern why men depart
To seek the Happy Isles!

That's what he thought about Auckland. I have no idea what he thought about Devonport. I remember very well, however, what I thought about Devonport. I thought it was the end of the world. That was a strange thing: we carry our geography with us, and wherever we are is the centre of our universe. But I, and other New Zealanders of my generation, felt acutely that we were on the edge, left out of it, and that the enormous, throbbing hubbub of the real world was far over the horizon to the north. As, indeed, it was: India became independent of the British Empire, and the empire began the awkward process of becoming a Commonwealth; Berlin was blockaded by the Russians, and fed by a continuous airlift by the British and American authorities; Western forces became bogged down in a brutal war in Korea, and Viet Cong soldiers encircled the French Foreign Legion at Dien Bien Phu; a king died and a young queen took over the British throne; there were bloody struggles for independence in Kenya, Cyprus, Aden and Malaya; the first tests of the hydrogen bomb were held at Bikini atoll; the Iron Curtain advanced across Europe; Britain, France and Israel tried to invade Egypt; the Russians invaded Hungary. We knew about these extraordinary events because we could hear radio broadcasts relayed from the BBC, there were newsreels at every cinema showing, and there were agency dispatches in the *New Zealand Herald*. But they seemed improbable, inexplicable and unreal.

Or perhaps it was Devonport that seemed unreal, because, at least in memory, nothing seemed to happen there at all, not even vandalism or car accidents. The circumstances that led me at nineteen to book a passage on an Italian passenger ship called the *Castel Felice* to Southampton via Singapore, Colombo, Aden, Suez and Naples seem barely worthy of remark. In those days, that was what young New Zealanders and Australians did. I rented a bedsitter in Earl's Court, then two rooms in Hampstead, and then a flat in Hull. I married and then successively purchased houses in east Hull, in east Kent, and in Kingston-upon-Thames before my family and I moved into the house at Hastings, situated at 50 degrees 51 minutes and so many seconds North and zero degrees 35 minutes and a bit more East.

Longitude zero sounds like a proper starting point, a place to head for, but this most assertive of timelines runs through France and Spain without bisecting any famous cities, and cuts across North Africa touching on Fez in Morocco before it slices into the blue Atlantic and stays there all the way to Antarctica. For such a historic meridian, it seems an uneventful place to be. Latitude 51 N however is much more promising. It crosses or passes close to Waterloo in Belgium, and Aachen that was once Aix-la-Chapelle, before skirting Cologne, Weimar and Dresden in Germany, and running just south of Wrocław in Poland that was once Breslau. These are all place names that had resonance, even fifty years ago in a wooden weatherboard house on the shores of a new-found land in another ocean, under a different set of stars.

It occurs to me, not for the first time, that even when I set off somewhere, I haven't left anywhere. All the places in which I have ever lived are still mine, in the sense that my memory contains them. They provide the theatrical backdrop to many of those intense moments that seem to have shaped and coloured my life, and their substance still intrudes in unbidden memories, old photographs and unexpected conversations. When we move from this present house – an increasing possibility at the time of writing – we will leave it, but it may not leave us. We head for new horizons, only to discover that we have already been there. We change our homes, but they stay with us. The story of our lives is a series of entries, many of them faded or crossed out, but still legible through the scratched pen marks: rather like the entries in an old address book.

TWO

The Town

The Man from Somewhere

In June 1652, smallpox being rife in London, the diarist John Evelyn left his wife and mother-in-law taking the waters at Tunbridge Wells, and headed for London. It was hot, he sent his manservant on ahead, and rode 'negligently under favour of the shade' till he got within three miles of Bromley, at a place called the Procession Oak. Suddenly, 'Two cut-throates started out, and striking with long staves at the horse, and taking hold of the reines threw me downe, took my sword, and haled me into a deepe thicket some quarter of a mile from the highway, where they might securely rob me, as they soone did.' Evelyn observed that 'it would teach me never to ride neere an hedge, since had I been in the mid-way they durst not have adventur'd on me; at which they cock'd their pistols, and told me they had long guns too, and were 14 companions'.

Evelyn lived, of course, to tell the tale. The story is a reminder that towns grew up as they did not because they could deliver jobs, public transport, banks, schools, recycling centres and planned housing: they grew up as they did to offer security to like-minded people in a menacing world. For much of European history, the countryside could be a dangerous place, home to vagabonds, beggars, fraudsters, robbers, smugglers, outlaws and freebooting soldiery. Kings reigned in capitals, but farther from the seats of power, their barons could do as they pleased, and often did; and pilgrims, merchants and craftsmen travelled in convoy, or hired an escort, or kept together as best they could, for their own protection.

The wildwood was a threatening environment. In 1285, the Statute of Winchester by Edward I required local landholders to clear the scrub, coppices and ditches for two hundred feet on either side of the highway between market towns, to prevent surprise attacks by highwaymen, footpads and robber bands. In such a world, European towns became places of safety for the people of the countryside: safe from outlaws, safe from the notional upholders of the feudal law, too. Perhaps because they had gates that could be locked at night and watched by daylight, towns became independent entities: communities of free citizens who could control trade and join guilds and plan cathedrals and repair battlements; could impose quarantine to keep leprosy or the plague at bay; could support hospitals and almshouses for the sick and elderly and build bath-houses and complain about sanitation; could levy tolls and impose building regulations; could decide to support a prince, or defy him; who could declare for the king or the Puritans in one country, or for the pope or the Cathars in another, or even declare neutrality, as Bremen did in the Thirty Years War.

Medieval towns, says the historian Lewis Mumford in *The Culture of the Cities* (1938), 'won the right to hold a regular market, the right to be subject to a special market law, the right to coin money and establish weights and measures, the right of citizens to be tried in their local courts and to bear arms in their own defence'. These were places that offered security and demanded loyalty; and each was different, with variant laws and customs, differing traditions and festivals, unique privileges and obligations, and unique penalties, too. Such towns might even, once they reached a certain level of commercial power and population density, grow into city-states or self-contained republics: Florence was one, and Venice another; Geneva a third, Strasbourg a fourth. Something of this sense of community, this sense of mutual support, survives in all local patriotism. A town was once, and in some ways still is, a place of belonging. You belong to Glasgow. Chicago is your kind of town. Maybe it's because you're a Londoner.

This sense of membership, of belonging, becomes in some ways a part of the identity that others confer upon us. Dickens's wonderful novel *Our Mutual Friend* announces one of its themes and introduces one of its protagonists in the second chapter with an idle dinner-party conversation about a capricious legacy that has befallen a man whose origins nobody can remember: Jamaica, Tobago, or … and it comes to the narrator at last, 'the country where they make the Cape Wine'. For conversational shorthand, however, the migrant in search of his inheritance has already become simply 'the man from somewhere'. Interestingly, although it is his life history, and his strange fortune, the bizarre codicil to a dusty legacy, that is compelling, to achieve even conversational reality, this person must first of all have a geographic location, however insubstantial: the man from somewhere. He must be placed, before his history can begin, and even though he is no longer there, he has departed, he has earned the preposition 'from', and his place of origin, however imprecise, is his first identification.

Towns confer an identity that grows from a sense of community. That word carries a strange burden, even in the original language. The French word *commune* in Latin was something that was common, in the sense of shared, for common use; it could also mean a community or state. *Communio*, according to my ancient *Latin Dictionary for Schools*, implied fellowship and mutual participation; but it also had a second meaning: to fortify on all sides, to barricade, to entrench. If you were a townsman, you were not a villein, not a feudal serf, not a bondsman, not tied to a manor or fated to till someone else's fields, you were not a beggar or a vagrant, a person with no rights, and with no fixed abode. Citizenship of a town authorised not just security and a certain fragile dignity, it also conferred identity, and this identity lives on in names, stories, plays, songs, psalms and ballads, in signatures, in public records, almost everywhere we choose to look, and sometimes we become so accustomed to the organic link between townsman and town that we barely notice the conjunction.

For example, when Jesus of Nazareth could no longer carry the cross on which he was to be crucified, Roman soldiers pressed a foreign visitor or tourist to help him. That is why we know about Simon of Cyrene: because three Evangelists, Matthew, Mark and Luke, all record the incident. After Jesus died – with, according to the Gospel of John, the Roman lettering *Iesvs Nazarenvs Rex Ivdaeorvm*, or Jesus the Nazarene, King of the Jews – above his head, a rich man called Joseph of Arimathea asked the Roman governor for his body. This story achieved its widest circulation once the Old and New Testaments had been printed in moveable type by Johannes Gutenberg. But even people who have never read the Gospels know the story, because episodes from the life and death of the Nazarene have been discussed by Augustine of Hippo and St Teresa of Avila and painted by Raphael of Urbino and Leonardo da Vinci. I can find no reference to Cyrene or Arimathea in the essays of Michel de Montaigne.

There is a clear theme here: everybody in this account so far is quite naturally identified by his or her place of origin, that is, the place from which they had come, and where they might normally have been found. In some cases, the identification is so sure and so natural that the name-bearers are sometimes referred to only by the places from which they had come. Michel Eyquem took the extended name de Montaigne from the château that his trader family had purchased in Perigord. His *essais* (literally, attempts) were composed there, and he is now known in the English-speaking literary world simply as Montaigne. Johannes Gensfleisch zu Laden zum Gutenberg's final surname comes not from his father but from the agricultural estate where his father lived. Vinci is a small Florentine commune or town in Tuscany that has been in existence for at least eight hundred years: so emphatic, however, is the fame of its most famous son that (once again in the English-speaking world) he is often and no doubt incorrectly referred to as da Vinci, as if it were a formal modern surname, like de Tocqueville. Nazareth is still known in Hebrew as Natz'rat and in Arabic as al-Nasira. It is an Israeli town with a large Arab and Christian population, and it is incontestably the home of the Jesus of the Gospels: his birth in Bethlehem is reported as an administrative accident during an episode of temporary urban overcrowding. Nazareth in Galilee dates from the Bronze Age: two thousand years ago it had a reputation of sorts among people elsewhere in the region because, in the Gospel of John, a certain Nathaniel asks rhetorically, 'Can good come from Nazareth?'

Arimathea, on the other hand, is described by Luke as 'a city of the Jews', but which city, and where it might be found, is not known. Biblical scholars have proposed several candidate locations, including the modern Palestinian town of Ramallah, on the West Bank, but for all the available evidence Joseph might as easily have come from Glastonbury in Somerset, a town much linked with his name. Cyrene is more easily identified: it was an ancient Greek colony – known to Herodotus – on the African shores of the Mediterranean, and part of Libya is still called Cyrenaica. Like Alexandria and other Greek cities, it had a Jewish population, and one imagines (because there is no evidence either way) that Simon was in Jerusalem for the

Passover, perhaps visiting friends and relatives (because Mark calls him 'the father of Alexander and Rufus'). These relatives might not have lived in Jerusalem, however, because Luke and Mark both describe him as 'coming from the country'. This phrase may have meant either the countryside, as in 'not the town', or the country where Cyrene lies. I do not know. We do know he didn't volunteer to help Jesus with his burden: all three Evangelists say he was compelled to help. The point is that all three choose to identify him by his place of origin. Of course they would. In a city and rural region that must have been home to hundreds of Simons there had to be additional identification: Simon the fisherman, Simon son of Jonas, Simon Magus, or Simon, the one from Cyrenaica.

Our towns become us, perhaps most of all when we are away from them. The John of Gaunt who with his dying words celebrates England in Shakespeare's *Richard II* ('This blessed plot, this earth, this realm, this England, This nurse, this teeming womb of royal kings'), was historically John of Ghent – he was born in that town in Flanders in 1340 – as well as Duke of Lancaster and King of Castile and Leon. Other people quite unconsciously tend to place us, a neutral act that is entirely to do with geography, rather than classification, as in 'put us in our place'.

Sometimes this works in reverse: an individual bequeaths his name to the town. Adelaide, Darwin, Melbourne, Brisbane and Sydney in Australia; Auckland, Wellington, Napier and Palmerston North in New Zealand; Vancouver, Halifax and Churchill in Canada; Alexandria in Egypt; and so on, were all named for individuals. These namings however were salutes and celebrations rather than claims of direct connection: in most cases the name-bearer never visited the namesake. There are rare cases in which the identities of town and first citizen seem difficult to separate: outside Geneva is the commune of

Ferney-Voltaire. It was simply called Fernex or Ferney until François-Marie Arouet, one of whose pen names was Voltaire, moved there in 1759.

Sometimes we salute the town unconsciously when we reach for the wardrobe, because towns dress us. The whole world has adopted denim, originally known as serge de Nîmes, after the town in southern France where it was traditionally made, and cambric takes its name from Cambrai in northern France. But islands, districts and townships have also given their names to a variety of clothing forms: Jersey, Guernsey and Arran for woolly knitting; Harris for a tweed weave; calico from the southern Indian city of Kozhikode once known, to the English at least, as Calicut; and so on. Towns give their name to products: Bakewell tart, Pontefract cake, Stilton cheese, spaghetti Bolognese, Chelsea bun, frankfurter and so on.

Towns also address us. Hastings is a common surname in the British Isles, Australia, New Zealand, Canada and the United States, and there isn't much doubt that as a surname, its origins lie in the town of Hastings. The suffix 'ing' once meant 'the people of', and one version of the town's history has it settled by an invading Saxon clan called the Hastingas, or sons of Haesten. There is a similar-sounding Norman word that implies swiftness, speed or haste, and Hastings is where the Norman invaders landed, and fought and defeated the English in 1066; but the settlement name of Hastings is a great deal older than that: the 'men of Hastings' get a separate mention in the story of Offa, king of Mercia, who seized Sussex in 771, and according to the *Anglo-Saxon Chronicle*, Danish marauders committed violence there in 1011. The town was the first Norman conquest in Britain, the site of the oldest Norman castle, an early gift of land by the conqueror to a loyal tribesman. There is an eleventh-century title of Baron Hastings: it was held by John, son of Henry de Hastings. Shakespeare evokes one of his descendants

in three of the history plays. '… ere I go,' says Edward IV, in
Henry VI Part Two,

> Hastings and Montague,
> Resolve my doubt. You twain, of all the rest,
> Are near to Warwick by blood and by alliance:
> Tell me if you love Warwick more than me?

Warwick, too, is a place that became inseparable from a name,
especially in Australia and New Zealand, two countries rich in
toponymic first names such as Warwick and Clyde.

So there are a number of levels at which we accept that our
identity derives from our place in the world. These connections
between identity and place seem the more natural because they
are timeworn: we are comfortable with a book title such as *The
Da Vinci Code*, but nobody, as far as I know, ever refers to a
painter called Urbino; he is either Raphael, or Raphael of
Urbino.

In a café in Cambridge, Massachusetts, a bartender once
hailed me as 'London' to bring me to the counter to collect my
hot salt-beef on rye: she knew the names of her regular custom-
ers, and by default she identified me by the place that I had said
I was from. I heard her later call out the word 'Ohio' for somebody else's double cheeseburger. It seemed, at the moment, quite
reasonable in the land in which movies could be named for the
Man from Laramie or a bandit called the Sundance Kid because
he once served a prison sentence in Sundance, Wyoming.

This identification with place and origin must have made
sense in an older world, too – one in which many people must
have stayed where they were born, and rarely travelled for more
than a day from their homes, for most of their lives. It is possible
to identify a very tentative modern trend in what might properly be called place-name christenings: Chelsea for a president's

daughter, Brooklyn for a footballer's son. But it remains hard to imagine Londoners being identified by registrars as 'Catford' or 'Peckham', or young bucks addressed as Milton Keynes. One feels that Henry de Hastings was plausibly so-called because he more or less owned the place, could deliver its rents, loyalty and fighting men to the king's cause, but most of all because he was another Henry, the one from Hastings. Back in Hastings, he would have seemed an outsider, because of the time he spent at court. But this connection of identity with place is a little tenuous and uncertain everywhere, and through all history, and so, on occasion is its reverse: the connection of place with identity.

I live in one of the most famous towns in the English-speaking world, but about this fame hangs a faint air of embarrassment, a whiff of inferiority complex. J. Manwaring Baines, a former local museum curator and author of *Historic Hastings*, a 1955 memoir, observes that 'it is rather remarkable that although so many noteworthy people have visited Hastings, few of its citizens have achieved national fame'. The town's name is linked forever with William the Conqueror, also known as William the Bastard and William, Duke of Normandy, and the Battle of Hastings in 1066; but William's cross-Channel invasion force neither landed at Hastings, nor fought there. William brought his troops ashore at Pevensey, and defeated the forces of England at a place now called Battle, miles to the north of Hastings. As an ancient town that grew swiftly in the seaside property boom that followed Regency investment in Brighton during and after the Napoleonic wars, Hastings has had its share of famous residents, but many of them stayed there only fleetingly. The pre-Raphaelite painter Dante Gabriel Rossetti brought the beautiful model Elizabeth Siddall to Hastings, and later married her there. Sir Arthur Wellesley, later Duke of Wellington, veteran of the eighteenth-century campaign to subdue India, was quartered there, and returned for a while during the war against Napoleon. Another soldier briefly billeted in Hastings and celebrated by a blue plaque was Sir John Moore, who died during the Peninsular War at Corunna in Spain and was immortalised by the poet Charles Wolfe in verses that children of my generation were required to learn by heart ('Not a drum was heard, not a funeral note, as his corse to the rampart we hurried …'). Byron swam off Hastings in August 1814, ate turbot there, walked the cliffs, and in a fit of pique threw an ink bottle out of a window. Keats visited Hastings, and Edward Lear painted there. A plaque commemorates Admiral Sir Cloudesley Shovell, who in 1707 sailed his squadron onto the rocks of the Scilly Isles, and perished with 1,400 other men. His connection with Hastings is that he 'in about 1700 took the opportunity whilst sailing past Hastings to visit the house reputedly lived in by his mother'. Titus Oates, a turbulent figure from the seventeenth century (dismissed as an Anglican priest for drunken blasphemy, discharged as a ship's chaplain for buggery), certainly lived in Hastings in his youth: he is now famous for producing fraudulent claims of a Popish plot against King Charles II, charges that ended with the murder and execution of a number of Catholics. Oates was later exposed as a liar, prosecuted for perjury, fined, pilloried, whipped, imprisoned for life and then a few years later released. The town's shingle beach, its fishing fleet, its long history and its picturesque streets made it a popular resort for painters, but two of the best-known are linked not with the Old Town, but the new town, on the western side of the West Hill. Not far from my house in Hastings is one of the temporary homes of Robert Noonan, painter, signwriter and decorator. Noonan achieved his fame under the pen name of Robert Tressell, author of the socialist classic *The Ragged Trousered Philanthropists*, in which Hastings achieved pseudonymous immortality as a place called Mugsborough. In fact, Noonan was a Dubliner who emigrated to South Africa and died in Liverpool: he lived only nine years in Hastings.

A little further on is another house linked with a painter. The resident was to become one of the world's most famous painted ladies. Anna McNeill Whistler, subject of a painting in the Louvre officially called *Arrangement in Grey and Black*, but known everywhere as *Whistler's Mother*, moved to Hastings in 1875 and died there in 1881. She too came from somewhere else: North Carolina in the old American South. Members of her parents' families kept slaves before the American Civil War; she married George Washington Whistler, an Indiana-born soldier who became an engineer, who built the first mile of passenger track in the US for the Baltimore and Ohio Railroad, and who invented the locomotive steam whistle. She moved to Russia, because her husband served the imperial court as a consulting engineer during the construction of the first main Russian railway line, the link between Moscow and St Petersburg. These were the tracks that carried the train that brought Tolstoy's heroine Anna Karenina to Moscow, to her fatal encounter with Vronsky, to public shame, unhappiness and finally death under the wheels of a locomotive. So Anna McNeill Whistler had already played, by marriage and location, an incidental role in the making of the cultural heritage of Europe, long before she came to England. She had five children: one of them was the painter James McNeill Whistler, and he composed her portrait in 1870, while she stayed at his London flat. Both Mrs Whistler and Tressell were conspicuous outsiders: that is, they were not Hastings people.

But I now realise that, in my twenty-three years in Hastings, I have probably known very few people who could claim to have been born in the town, or even to have spent most of their lives there. Of the houses that border our own house and garden, more than half have changed hands not just once, but several

times in those twenty-three years. Most of my immediate neighbours had moved to the town, as we had done, from some other
place: London, Kent, rural Sussex, Surrey, the English Midlands,
the West Country, Scotland, Wales, Ireland, France, Poland,
South Africa, the United States and Australia. There is seldom a
single reason for settling in one town rather than another: family
connections and job opportunities play a role; so too does the
cost of housing, and for most of my twenty-three years there,
Hastings was conspicuously more affordable than many other
south coast localities. Physical beauty must also have been part
of the lure.

Hastings began as a seaport, one of the Cinque Ports, a
confederation of towns required by a Royal Charter of 1155 to
maintain ships for a royal navy, should such a need arise. In
return, Edward I gave Cinque Port townsmen the right to bring
goods into the country from abroad without paying import
duties, a right that seems to have been very unwillingly surrendered sometime in the reign of Edward III, and smuggling
remained a significant source of income well into the nineteenth
century. Hastings's role as a naval and military base was relatively short-lived. What had once been a secure harbour was in
the thirteenth century swept away by storms and coastal erosion,
but the town survived as a home to fishermen, smugglers and
traders, and it grew up around a small sheltered inlet in a narrow
valley and then spread inland between two steep ridges called
simply the West Hill and the East Hill, along the course of a
stream called the Bourne: this is now submerged beneath a high
road still called The Bourne. The town experienced the usual
buffets of history, along with plague, fire, assault and property
speculation followed by slump. It remained well known without
ever becoming very fashionable, and much visited without ever
becoming rich. In the course of the last five centuries, houses
were renewed or restored, propped up and extended, built over

and upstaged, in ways that would now give town planners the vapours. Old shops, warehouses and alehouses would become homes, and then turn back into businesses, or restaurants, or studios, or antique shops, and then morph back into accommodation again, while still retaining the paraphernalia of commerce about them.

Hastings Old Town was where I first came across the phenomenon of a 'flying freehold', in which some long-gone property-owner might add an extra attic room or two to his house by building it over the roof of his immediate neighbour, and then leave subsequent generations of lawyers, tenants and freeholders to sort out the consequences for themselves. In one case, access to an eighteenth-century hillside cottage was only possible through the garden of the house immediately below it; in another, rows of little terraced cottages could be approached only by a footpath or a flight of steps; in a third, somebody's little front garden and terrace sat on top of somebody else's lock-up garage. Whole streets of properties could be quaint, picturesque, charming and barely habitable, so small were the staircases, so cramped the rooms.

The appeal of old Hastings had a great deal to do with its long-term impoverishment, its place at the bottom of the property pecking order: old Georgian tiles and Welsh slates stayed where they had been put because owners could not afford to modernise; sash windows continued to rattle in the gales because unplasticised polyvinyl-chloride-coated double glazing was too expensive; householders had quite enough to manage, coping with dry rot, damp, death watch beetle, subsidence, defective drains and the notorious impact of salt-laden wind and rain on the property's paintwork and timber. But for many visitors, the unexpectedness of some of the houses was its own reward; for others it was the view. Someone standing on the West Hill of Hastings could see, looking due south, a long sward of green punctuated by people walking dogs, children playing with kites, and little knots of picnickers, and beyond them the broken stones of a not quite levelled Norman castle, and beyond that the English Channel, the colour of pewter on an overcast day, a sheet of silver and azure when the Sun shone. A visitor who turned slightly to the east could see a tiny fishing harbour protected by a small breakwater, a row of broad-beamed, sturdy fishing boats hauled up each day onto the pebble beach, a collection of tall net-drying sheds protected by coatings of creosote or pitch, a fairground, a disorderly jumble of attractive roofs and house fronts on both sides of the valley, huddled together either for economy or security, with small and often steep gardens.

Within view would also be two fine stone churches, and a handful of conspicuously handsome but not especially large properties. The townscape would be framed by woodland and terraces of allotments on the steep western hillside, but on the eastern hill, the houses would be crammed together until, just about halfway up the slope, they would stop, as if restrained by a regal hand that had drawn a line with a ruler, and beyond that the observer would see only the grassland and scrub and forest of a country park, ablaze with gorse and hawthorn in spring, and arrested along the coast by steep sandstone cliffs.

This extraordinary view – you could never get tired of it, which was just as well because it was almost inescapable – belonged to everybody, and was divorced from all other privilege. Some of the poorest families, housed in an estate along the slope of the West Hill, probably had the best outlook in the whole of southern England: a panorama framed by two headlands, a valley crammed with picture-postcard roofs, a beach and seaport that had been a magnet for painters for two centuries, and then beyond it, just the sea and the sky.

This prodigal exposure to beauty came at a small but inexorable cost: prodigal exposure to wet and cold when the clouds gathered and the sea began to pound the coast and shrieking winds from the Bay of Biscay, or from the Baltic, would drive rain almost horizontally into the hillsides. People on the sea front would find passage perilous, as waves swept up the shingle to thunder against the stone walls above the beach, and explode in a shrapnel burst of freezing salt spray against the buildings on the landward side of the road.

The hills on either side of the Old Town were so steep that some of the access was by stepped passages, and for short periods during the more severe assaults of snow and ice, neither taxis nor buses would tackle the climb up the West Hill. At such inclement times, the short walk across the open green before the more sheltered descent to the shops, cinema and railway station was a headache of a different kind: the howling winds could turn an umbrella inside out, drive the rain down the lining of the strongest waterproof and chill the exposed cheekbones, sinuses and the temples to temperatures so agonising that the sufferer, once under shelter and back in the warmth, required a few minutes of privacy to sob and gasp with the pain that arrived as the numbness ebbed.

Most people, most of the time, accepted such discomforts philosophically, as the obvious and unavoidable reciprocal of the rewards of seaside life. These included fish caught that morning, landed before noon and sold at shops that backed onto the fishing harbour, or freshly preserved in a smokehouse across the road; a more than usually liberal supply of inns, bars, cafés, restaurants, bakeries and fish and chip shops aimed at the tourist trade but sufficiently dependent on patronage from the townspeople to keep the prices keen; and enough antique, second-hand and knick-knack shops to guarantee – for the purchaser prepared to plod a sufficient distance – almost any clothing, furniture, instrument, implement or distraction desired.

Seaside communities share a set of characteristics that would define them as a species in the taxonomy of township: for one thing, they smell of the sea, parked cars are splattered with seagull excrement and the buildings nearest the front tend to look either freshly painted or faded and weatherbeaten, because these are the only two states possible on a southern coast exposed to prevailing winds. The seawall, promenade or coastal defences provide one obvious limit to expansion, and ambitious town planners hoping to exploit the holiday trade long ago zoned conspicuous areas of parkland to enhance amenity, and at the same time took steps to preserve access to the open countryside, so light industry, warehouses and superstores tend to accrete asymmetrically in one direction along a coast road and on the high terrain inland from the newest developments. All such towns have seen better days, and hotels and mansion blocks once built by speculators to attract the regular patronage of gentry and aristocracy have in some cases become temporary and uncomfortable accommodation for the dispossessed: refugees and asylum seekers from a dozen civil wars in Europe, Africa and Asia; and the unemployed, divorced, rejected, alcoholic, drug-addicted and depressed from the cities to the north. The south coast towns embrace pockets of considerable poverty within bigger settlements of only modest wealth, but the communities of have and have-not seem to muddle along peacefully most of the time. The areas nearest the promenade, pier and beach tend to become no man's lands of peeling hotels, bars, amusement arcades, small ethnic restaurants, fish and chip shops, sweetshops, ice-cream parlours and kiosks specialising in souvenirs, ethnic jewellery, crass postcards and silly objects. In high summer and good weather, the beachfront roads and pavements are crowded with cars, cyclists, roller skaters, joggers, strollers, buskers and gangs of teenagers; the numbers are intermittently swollen by chartered busloads of day trippers, some of them families from south or west London out for a picnic in the fresh air; some of them visitors from Belgium or Holland. Near the close of the day, the numbers diminish, and the seagulls reclaim the shore, to peck at discarded and rejected kebabs, half-eaten pizzas and discarded boxes of chips.

Although such things define the seaside town as a species, each English south coast settlement retains a distinctive individuality, and could not – except for limited aspects within the 'tripper zone' that they all share – be mistaken for any other seaside town. Eastbourne, on the western side of the Pevensey Levels, has a spacious and gracious old town of flint and brick buildings served by brick pavements of the sort I have seen nowhere else, and an extended and cultivated promenade that gets wilder and more beautiful the further westward it is from the town's heart. Hastings town comes almost to a dead halt around a crush of little buildings opposite the fishing harbour, at the foot of the East Cliffs. Its promenade begins beyond a seafront fairground and extends gracefully enough beyond the town's pier, all but destroyed by fire, and along in front of a succession of handsome buildings all the way past St Leonards to a place known as Bo-Peep, where the attempt at formal seaside identity collapses into a clutter of light industrial buildings and small houses, beach huts and boat sheds and a long shingle walk to a huge, detached shopping development that announces the start of Bexhill-on-Sea. This southern perimeter is the baseline for a bulging triangle of suburban development that extends north to a road called The Ridge, beyond which the countryside begins. At the 2001 census, this area was home to 85,029 people, with a mean age of 39.6 years, of whom more than 94.5 per cent were born in the United Kingdom, 67.4 per cent of whom identified themselves as Christian, and only 2 per cent of other faiths.

* * *

In 2001, I and my family were of this number: now, in this addendum to a chapter that has taken more than a year to complete, we are not. We have moved across the Pevensey Levels to Eastbourne; another town that grew up around a bourne or stream, another town mentioned in the Domesday Book, a town in its own different way as distinctive as Hastings, and to a house in no way as distinctive as the one we maintained in Hastings. The reasons for the move are not important: what matters is that we felt no great wrench, no dislocation and no sense of loss as we made it. The twenty-three years that I spent living in Hastings was the longest period of my life in any one settlement, let alone one house. If we count by years of residence, then that would make me a Hastings man. But it does not. The towns of Kent and Sussex, of Essex and Surrey and Hertfordshire and Buckinghamshire, have to some extent struck a bargain with the national capital, London. You provide the employment, and we will provide the people; you provide the income, and we will furnish the beds, the county towns might have said. The distinctive autonomy, the panoply of entirely local legislation, custom and practice that distinguished one community from another, began to disappear centuries ago as central government exerted increasing authority. The arrival of the railways did the rest: a million or two households in the south-east of England depend on earnings paid by London-based employers, and places such as Kent and Sussex are simply part of a larger commune called the commuter belt. Townspeople from Hastings – and from Tunbridge Wells and Ashford and Lewes and so on – began to get used to spending hours of each day neither at home nor at work, as citizens of nowhere in particular, travelling at an average of fifty miles an hour through landscapes at which they have long since ceased to look.

Once a town becomes a dormitory suburb of a metropolis, a place of convenience and affordability, it loses some of its character. Economic policies now encourage the movement of capital and labour, but it is important to remember that they always did. Towns welcomed merchants and merchandise; after the great plague of 1348, which is supposed to have killed more than one person in three in Europe, the survivors also welcomed any craftsman or skilled labour that came their way. For precisely the same reasons that medieval settlements grew into fortified towns, each with its own unique character, so this uniqueness began to dissipate. People have always been prepared to move home, to go somewhere else in the hope of a new and perhaps a better life, a greater challenge, a more thrilling vocation, and this departure is the thing that confers some of the substance of identity. Jesus the carpenter's son became Jesus of Nazareth when he left Nazareth, first for the countryside and then for Jerusalem. At home in Vinci, Leonardo was simply the son of Messer Piero; he became Leonardo da Vinci when he moved to Milan, Florence, Venice and Rome. But bureaucracy now confers upon us other, more precise forms of identification: we have formal surnames, registered at birth and recorded on a passport. These surnames are sometimes place names – Norton, Barton, Sutton all date back to old English locations – but they define lineage rather than place. We acquire tax and National Insurance numbers, and must be identified by our dates of birth. The town exists as a postal address, but it could, in Britain, be supplanted by that unfeeling and relatively recent precision tool, the postcode.

Most of us are uprooted, detached and replanted. The Duke of Devonshire has his historic home in Derbyshire. The Prince of Wales does not live in Wales: he may more easily be found in Scotland, Norfolk, Buckinghamshire, London or Gloucestershire. And this writer, who made a home in Hastings for almost a third of a lifetime, has somewhat ruefully discovered that he is not a Hastings man. He has become a man from somewhere. He is a

United Kingdom citizen; he has made his home in England for almost fifty years; he loves Sussex; he is married to a Londoner; his son is a Yorkshireman by birth, and his daughter a Maid of Kent; he is a monoglot Francophile and Russophile and emphatically European in his political sympathies; but he cannot shake off an awareness, a cast of mind, a substrate of identity that set in his teenage years, under bright skies in which the Sun was always in the north, beer was always served cold, roads were covered with loose metal, houses had verandas, trees were evergreen, Christmas Day was hot and children ran to the beach in bare feet.

To be aware of this is not to wish to go back. We cannot go back to what was, because that too has altered, and in some ways vanished. However assimilated one tries to be, however aware of the history, geography and romance of the new location, one remains a stranger in one's adopted country, and at the same time one becomes a stranger in one's place of birth. That may not present a disadvantage: people who never leave a place never really understand where they are. Those of us who change places, who migrate, perhaps get the best of both worlds. It was a commonplace of 1950s fiction that young people set off to find themselves: to find out who they really were. When you migrate, you find out who you are not; you can be at ease, and secure, and part of a community, but you also discover that you are never quite at home.

THREE

The County

A Piece of Chalk

Underfoot, the turf is short, rabbit-bitten and studded with tiny flowers with names that read like fragments of an unassembled poem: eyebright, silverweed and bird's-foot trefoil; wild thyme, yellow wort and scarlet pimpernel. To the right and left, nearer the woods, and standing much higher, are more wildflowers: hemp agrimony, self heal, woundwort, centaury, rosebay willow herb, wild parsnip, viper's bugloss and thistles that smell like spilt honey. Among them flutter the butterflies: my wife, the expert, identifies comma and peacock, red admiral, large and small tortoiseshell.

We are deep in the Cretaceous, in a shallow valley or bottom between two of the Sussex Downs, and everything about me is a lesson in the making of a county. The rock underneath our feet is a soft, porous form of calcium carbonate called chalk, and it extends into the Earth's crust for hundreds of metres. This chalk represents ancient life: each microscopic brick in this prodigious earthwork was once a tiny little planktonic alga called a coccosphere that flourished in the hot, sunlit seas that marked the Cretaceous period, which lasted for about eighty million years and ended sixty-five million years ago. Creatures similar to these still live in the tropical waters, well below the surface, but within the region that light can penetrate. They still use sunlight and the weaponry of photosynthesis to build tissue by absorbing carbon dioxide dissolved in seawater; they still bloom, replicate and die, and their tiny remains still fall slowly through the darkening waters, to settle on the deep seabed and become part of the abyssal ooze that cloaks the hard basalt of the ocean floor. The remains – bones seems the wrong word, because a coccosphere is a little assembly of tiny scales or plates, shaped into an orb – are called coccoliths. It isn't hard to imagine a great, warm ocean under sunny skies; it isn't hard to imagine the light flickering diffusely through the relatively still water far beneath the waves; it isn't hard to imagine microscopic plants slowly forming, hijacking energy from this filtered sunlight to construct themselves, molecules at a time, from material dissolved in the waters around; it isn't hard to imagine many of them being eaten by zooplankton or consumed by fish which would in their turn be devoured by larger predators. It isn't hard to imagine that enough would survive to die naturally, and begin the long, slow fall into the darkness below.

It isn't hard to imagine any of these things; it is however hard to imagine such things happening for eight, or eighteen, or eighty million years. People who tick off generations in twenty-five-year spans, who expect to die at around the time their grandchildren are ready to reproduce, who count national history in centuries, who measure the history of civilisation in millennia, simply cannot absorb the immensity of a time span measured in tens of millions of years.

Algal plankton are not visible to the naked eye. Their mortal remains are even smaller. What kind of dusting would invisible skeletons leave on the ocean bottom, and how fast would this dust assemble? Under the pressure of more and more invisible skeletons raining down, to create at first a protective layer, and then a blanket, and then another blanket, and then the submarine equivalent of an eiderdown, and then another eiderdown, and then a mattress, at what stage would the cumulative, smothering weight finally crush the bottom-most layers of this soft, yielding bed into rock? How long would it take to make what now represents a centimetre of chalk, a cylinder of which a teacher might once have used to inscribe history on a blackboard, or in exasperation might once have thrown at an inattentive pupil (something that teachers used to do)?

Suppose it took a thousand years to make one centimetre of this undisturbed, impacted submarine dust. In 100,000 years, there would be a metre of chalk. In a million years, there would be ten metres of chalk. The chalk of the Sussex Downs extends for five hundred metres, and nobody can possibly know how much of this chalk has already been eroded: washed or blown or scraped away by rain, wind and ice, or scoured away by advancing and receding seas. The white cliffs of Dover in Kent and the Seven Sisters of Sussex are a work in progress: in the progress of demolition rather than construction, as the Channel tides and tempestuous waves driven by seasonal gales tear away at their basements, and bring down the fabric above. But as long as they endure, the cliffs stand as testament to warm seas, sunny skies and a world occupied by strange creatures.

The Cretaceous period was one in which there was almost no ice or snow, except on the highest mountains. Forests grew in Antarctica, and on the north slopes of Alaska. Dinosaurs grazed, hunted, reproduced and died far within the Antarctic Circle. Sea levels were far higher than today: sometimes hundreds of metres higher. The chalk exposed in the iconic white cliffs of south-eastern England is only a small fraction of all the chalk there is to be found: there is chalk beneath, and occasionally exposed at the surface of, Yorkshire, and East Anglia, and France, and Germany, and much further, too. Darwin's bulldog, Thomas Henry Huxley, in a famous essay on 'A Piece of Chalk', pointed out that it runs through Denmark and Central Europe, and extends southward to North Africa; while eastward, it appears in the Crimea and in Syria, and may be traced as far as the shores of the Sea of Aral, in Central Asia. If all the points at which true chalk occurs were circumscribed, they would lie within an irregular oval about three thousand miles in long diameter – the area of which would be as great as that of Europe, and would many times exceed that of the largest existing inland sea – the Mediterranean.

But this vast bedrock-in-the-making, this shelf of calcareous substrate, cannot have been submerged for the whole of the Cretaceous. Below the chalk lies greensand from the same geological period, a muddy sediment that speaks of entirely different conditions: of a huge, wandering delta landscape, of marsh, meanders, oxbow lakes and seasonally flooding rivers that would dry up and exist only as trickles, built up by silt from the mountains further inland, and then occasionally swept away again as the sea invaded, leaving its signature in the form of raised beaches.

These beds, too, extend hundreds of metres downwards: their rock is now the Weald of Kent and Sussex, and they represent a long overture, a prelude, a period of marsh and mudflat, water meadow and tidal estuary, beach, dune and delta. Through this well-watered, fertile silt stalked iguanodon and baryonyx, and above them flew pterosaurs. Elsewhere in the world, Tyrannosaurus rex was already on the prowl; ichthyosaurs and plesiosaurs hunted in the shallow seas. So the Cretaceous period is marked by episodes of high and low water, of violence and movement. The modern world had begun to take shape: the Atlantic had begun to open, and Africa had begun to close in upon what would become the Mediterranean, to start pushing up the Alps. The extended ripples of that impact would gently uplift the chalk and shape the contours that would one day become the Downs, and the sediments that would form the Weald. Western Europe might be partly or entirely submerged, but at intervals fragments of Britain and Brittany must have been visible high above the waves: Wales was a rocky island, and so was something that would one day become Cornwall. The Pennines stood proud, and there was once an East Anglian massif – yes, marvel at the mountains of Norfolk, now at sea level and sooner or later to be submerged – that towered above the submerged Wessex basin. This exposed rock was sheathed not just with the implacable green of fern and cycad and moss: it sported splashes of bright colour. Vegetation had begun to change, and with it the fauna of fields and forests.

The ancestors of modern butterflies and moths, the first ants, the first aphids, grasshoppers and gall wasps appeared during the Cretaceous, and so did the first flowering plants. The grasses will not emerge for aeons, but petals, pistils and pollen have begun to evolve and flowering plants to speciate in precise step with a new generation of pollinators, predators and scavengers. Chalk flowers such as eyebright and scarlet pimpernel have their roots in the Cretaceous, metaphorically as well as literally, and the comma butterfly is a kind of lepidopteran punctuation mark in a long story that begins with the chalk.

But this accidental epic contains another narrative, one from which the history of human science and technology grows. At intervals in the chalk are puzzling inclusions called flints. Chalk is soft, flint is hard; a lump of chalk can be of any size or shape, but flints are knobbly, uneven things usually measured in centimetres or tens of centimetres. The exterior of flint is white: it takes its colouring from the carbonate of lime in which it is found. The interior however is dark: it is a concretion of silicon dioxide, also known as quartz, and also known as silica. There is no satisfactory explanation for the existence of hard nodules of flint in soft chalk, but the long-standing conjecture is that these too begin with submarine biology: they could have been formed from the siliceous remains of Cretaceous sponges that once grew in clumps on the seafloor. Flints are exposed with every ploughing, and they became the raw material for the walls, barns and homesteads of old settlements in the Sussex Downs.

Flints also provided western Europe's first systematically exploited cutting-edge technology, its first tool of mass production, the Neolithic equivalent of the Swiss army knife. Bang a flint with something very hard, and you flake a fragment from it. Keep on doing that, and you can turn a nodule of flint into an axe, an adze, an awl, a chisel, a knife, an arrowhead, a spear point, a sickle, a razor, an item of barter or even a tool for mining yet more flints. Five or six thousand years ago, late-Stone-Age humans systematically dug a series of shafts and galleries in the chalk near Spiennes in Belgium, and excavated on an industrial scale huge quantities of flints that could be worked and then traded for goods from other locations. Simply to work the mines, get the product to the surface and then exploit its potential value, these Neolithic entrepreneurs had to be aware of the market pressures of supply and demand, the principles of sustained cooperative endeavour, the basic demands of health and safety management, the logic of shared income and the role of specialist craftsmanship in the wider commonwealth; of the notion of apprenticeship and education; and in addition the proper design and deployment of pit props, the planned removal of spoil and other mining requirements. All modern technology and business theory starts from the silica chip. One property of flint is that if struck with metal it will fire sparks: flint was the first portable, all-purpose firelighter, and thousands of years later would become the basis of the flintlock musket.

Silica in the form of sand became the basis for glass; it also became a bulk component of ceramic and of concrete. Out of glass, craftsmen ground the first lenses and assembled the first telescopes and microscopes, prepared the first prisms and made the first artificial rainbows from a beam of white light. The sciences of astronomy and navigation, of microbiology and the germ theory of disease, of spectroscopy and atomic theory, all begin with the exploitation of silicon dioxide. So a silica chip, first flaked from flint by an unknown hominid more than a million years ago, was the beginning of all science, and all technology; it was the beginning of the exploration of space and time, and of the tissue of life itself.

Flint wasn't the only agent of Palaeolithic cutting-edge technology: the earliest tool users exploited obsidian, and basalt, and bone and antler and shell, and even greenstone. But flint turned out to be the most versatile: so easily worked that tools could be made, used and abandoned; so easily found on or near the surface that hunting parties would make detours to known outcrops of chalk and flint to renew their weaponry again and again; so reliable that the tribesmen who camped in Belgium more than five thousand years ago considered it worth their while to establish a pithead, and a factory site, and to exploit the same resource for generations. And all this potential power was deposited and fashioned by the prevailing conditions deep in the oceans of the Cretaceous period, eighty or a hundred million years ago.

It seems presumptuous to claim that human ingenuity, technology and cooperation were driven by the discovery of flint; but it might be that without a reliable supply of superior toolmaking material early on in the story, without, so to speak, that extra edge, *Homo sapiens* might not have survived. The first anatomically modern humans are, genetically, so closely related that geneticists suspect that there might have been a population crash 70,000 years ago, leaving only small band of survivors to engender all the billions who now command the planet's resources. In a touch-and-go world, who can be sure what tipped the balance towards survival? Our species might have disappeared in the way that *Homo heidelbergensis, Homo erectus, Homo neanderthalensis* all did, the last of them around 30,000 years ago. So just as chalk and flint are part of the making of Sussex, chalk and the flint that it preserves are part of the making of human history.

This history begins, in Sussex, about 4000 BC. There were human settlers in the ancient wildwood long before that, but – I learn from Oliver Rackham's scholarly work *The History of the Countryside* – around 4000 BC new people arrived, bringing with them 'those crops, animals and weeds which constitute agriculture. They immediately set about converting Britain to an imitation of the dry, open steppes of the Near East, in which agriculture had begun.' Perhaps the innovators were invaders, perhaps some of the existing population simply imported new strategies. The debate continues. But the wildwood they cleared was not primeval – before the ice retreated around 11,000 BC, northern Britain would have been a very large icicle, and southern England a stretch of tundra – but across what would become Sussex and other south-coast counties must have certainly been a wonderful mix of lime, oak and elm, left more or less alone to grow to enormous stature. And the people who cleared the land may never have seen or known about the Near East: it is enough that their ideas, their technology, their management of the countryside began there. By the Iron Age, dated from the sixth century BC, much of the wildwood had gone, and there were secondary woods full of coppiced hazel and other timber grown for special uses.

The Romans occupied Sussex, but left only one famous road between London and Chichester, famous in its Saxon name of Stane Street. It is not, however, Britain's only Stane Street: there is another one that runs between St Albans and Colchester. They built a fort at Pevensey, and called it *Anderitum*, and at some time in the fourth century established a Roman commander called the *comes litoris Saxonici*, the wonderfully-named Count of the Saxon Shore, which alone suggests that Roman Britain was already under repeated assault. But this commander defended an area far larger than modern Sussex, and the name Sussex itself dates from the Kingdom of Sussex, or South Saxons, that began to form after the sacking of *Anderitum* or *Anderida* in AD 491, and Sussex anyway became absorbed into Wessex, and was then attacked and held for a while by the Danes, and Normans, also from across the Channel, had already set up in business in east Sussex many years before the formal invasion by Duke William in 1066.

Under the Normans, the management of Sussex, uniquely, was divided into rapes: initially Arundel, Bramber, Lewes, Pevensey and Hastings, each with a castle, a lord and a waterway. Chichester was added later. The use of an Old English word for a Norman administrative arrangement has provoked the historians and encyclopaedia compilers into wondering if these divisions had already existed when William came. But the act of division, or the formal exploitation of already-existing divisions, also suggested a sensible precaution on the part of William. If he could invade the Sussex coast and seize a kingdom, then so could somebody else, which is perhaps why his half-brother, and a son-in-law, and other reliable men, were awarded military command and control of the fateful shore.

There are no meaningful maps from the period, but the Sussex described in laconic detail in the Domesday Book of 1086 (Ditchling: 'King Edward held it. It never paid tax. Before 1066 it answered for 46 hides; when acquired only 42 hides; the others were in the Count of Mortain's Rape, and 6 woods which belonged to the head of the manor ...') seems to have been in area and boundary much as it is now. Alciston is described ('The Abbot of St Martins holds Alsistone from the King. Young Alnoth held it from King Edward.') Rye, which belonged to the Abbot of Fécamp in Normandy even under the reign of Edward the Confessor, had '5 churches which pay 64 shillings; 100 salt houses at £8 15s, meadow 7 acres; woodlands, two pigs from pasturage'. The towns, villages and hamlets along the footpaths and bridleways of Sussex that we know well all get a mention: Alfriston and Bexhill, Exceat and East Dean; Bodiam, Jevington and Herstmonceux; Netherfield and Wilmington and Wannock, and Rodmell where Virginia Woolf drowned herself ('Ramelle/Redmelle: William de Warenne, formerly Earl Harold, 11 salt houses, 4000 herrings'), have all existed for at least a thousand years (because the Domesday Book records ownership, entitlement and yield from before the Conquest as well as at the date of compilation), and probably a great deal longer.

There should be nothing surprising in the discovery that the settled landscape has changed so little in the last thousand years – fifteen or twenty lifespans, fifteen or twenty episodes of family life in which three generations could meet and feast in the same house – but somehow it is surprising. We tend through our education or our experience or our preferred reading to pursue intimacy with one period of history rather than another, and I cannot successfully imagine in any detail the England of the Conquest or the medieval centuries. I know a bit more about the England of Henry VIII and Bloody Mary and Elizabeth I, but it has always seemed an alien place to me: another country, where poets brawled and courtiers plotted and kings warred and landowners went to sea and explored new worlds and then came back and were put to death; where the devout were tortured and then incinerated for believing in the same God while following a different ritual of celebration; where Sir Richard Grenville could take on a whole armada, squander the lives of his seamen, and be hailed as a hero for it, rather than condemned as a pugnacious madman.

So the landscape of human ambition was different. The physical countryside beyond the cities seems to have remained throughout the centuries remarkably the same: the houses have changed, been rebuilt, extended and improved, again and again, but the terrain on which those houses stand seems barely to have altered. I stress the word *seems*. In 1586, William Camden visited Sussex, raced through its Roman past, briefly rehearsed its South Saxon origins, and then began to describe the county that I think I can see about me every time I go for a walk:

> ... it hath few harbours by reason that the sea is dangerous for shelves, and therefore rough and troublous, the shore also it selfe full of rocks, and the South-west wind doth tyrannize thereon, casting up beach infinitely. The sea coast of this countrie hath greene hils on it mounting to a greater height, called the Downes, which, because they stand upon a fat chalke or kind of marle, yeeldeth corne aboundantly. The middle tract, garnished with medowes, pastures, corne-fields, and growes [groves] maketh a very lovely shew ...

In fact, Camden confirms that Sussex has certainly changed: the 'rough and troublous' seas that he records have torn away at the downland cliffs, and drastically rearranged the levels of Romney and Pevensey, have left the port and harbour of Rye and the castle of Bodiam stranded miles from the water's edge, and have swept settlements away altogether. Camden describes Sussex:

> Full of iron mines it is in sundry places, where for the making and fining whereof there bee furnaces on every side, and a huge deale of wood is yearely spent, to which purpose divers brookes

in many places are brought to runne in one chanell, and sundry medowes turned into poles and waters, that they might bee of power sufficient to drive hammer milles, which beating upon the iron resound all over the places adjoyning.

Daniel Defoe, when he returned 150 years later, found Hastings barely worth a mention, and he observed that Rye

would flourish again, if her harbour, which was once able to receive the royal navy, cou'd be restor'd; but as it is, the bar is so loaded with sand cast up by the sea, that ships of 200 tun chuse to ride it out under Dengey or Beachy, tho' with the greatest danger, rather than to run the hazard of going into Rye for shelter.

Dengey and Beachy must be Dungeness Point and Beachy Head. Defoe too was interested in the heavy industry of Sussex: 'I had the curiosity to see the great foundaries, or iron-works, which are in this county, and where they are carry'd on at such a prodigious expence of wood, that even in a country almost all over-run with timber, they begin to complain of the consuming it for those furnaces, and leaving the next age to want timber for building their navies.'

The iron mines, furnaces and forges have gone. So in fact this sense of permanence, this feeling of enduring history, is a tease: landscapes alter. Humans make their mark and then in a generation or two the marks are erased: the evidence of what has once been may be visible enough to archaeologists, ecologists and topographers, to trained eyes, but to most of us the countryside we see is timeless: because it is there, and because it looks just so – a little pathway through woodland, a small clearing among the beech and Scots pine that is ablaze with rampion and bladder campion and stitchwort and rosebay willow herb – we find it easy to believe it has always looked so. A second inspection reveals that the woodland is in fact an old hedge that has been neglected for long enough that its hawthorn, elm and beech constituents have grown to full height, and its canopy has closed over the path. The beech and Scots pine trunks around the woodland clearing at closer examination are seen to stand in rows; the pines were planted, and many of them replaced after the first thinning with beech seedlings; the foresters have been here, and left a little place in the sun for the wildflowers, but already new saplings are pushing their way above the shrubs. The countryside tends to be seen as humans wish it to be. Anthropocentrics all, we see the landscape from our point of view, and even the entity we call the beauty of the wilderness is simply a happy arrangement of high ground and valley, glacier and river, forest and sky, that fits the unconscious frame of reference that we have for beauty: nature builds the structures, but we provide the composition.

Human appreciation changes too. Huxley, when contemplating the geology of Hampshire, Sussex and Kent, marvelled at the grandeur of the white cliffs of the Channel coast but patronisingly remarked that 'the undulating downs and rounded coombs, covered with sweet-grassed turf, of our inland chalk country, have a peacefully domestic and mutton-suggesting prettiness, but can hardly be called either grand or beautiful'. The inland downs between Brighton and Eastbourne now seem to me to be one of the grandeurs of the world, precisely because their rounded, comely perfection – Kipling called them the 'blunt, bow-headed, whale-backed Downs' – has been shaped by sheep. The Romney Breeders Association claims that the Romney sheep, the handsome, fleecy heavyweight of the Romney marshes, was the foundation of the great wool and mutton industry of New Zealand, but the other breed of sheep that we knew in New Zealand was the somewhat smaller, and perhaps more prettily mutton-suggesting Southdown, the basis (I am reminded by the Southdown Sheep Society) for Canterbury lamb, that is, lamb from the Canterbury plains of the South Island of New Zealand. This too takes its name from its place of origin, its first forwarding address, and this too is an old breed, one that would graze and manure downland pastures and enrich them enough for a wheat crop the following year.

The other great shaper of the downland turf, the animal that in some places nibbles the grassland so closely that it resembles a billiard cloth, is the rabbit. This little high-yield lunch on four legs, once a hardy native of the Mediterranean shores, was probably introduced by the Normans, as a commercial protein source. And, as William Camden so rightly pointed out, the Downs still yieldeth corn abundantly: each new horizon at the highest point of each down delivers yet another extraordinary panorama of smoothly rounded, closely-grazed coomb and valley; yet another little set of stone enclosures, barns or lambing pens; yet another stand of managed forest or – nearer the sea – windswept scrubland; and between these, rising beyond the flint walls and barns of the farmhouses, are huge fields of spring and winter wheat, barley, and sometimes even oats. There are dairy cattle in water meadows and floodplains of meandering downland rivers such as the Cuckmere; there are fields of pasture for horses near the villages; there are occasional crops of maize and oilseed rape and linseed; and there are hay meadows; but the Downs seem to have been shaped by sheep, just as surely as the commerce of England, after the Norman Conquest, was shaped by the export of wool to the Hanseatic cities across the North Sea; and just as surely as the economy of my own birthplace, New Zealand, was shaped by strains of sheep that over the centuries had adapted to the downs and weald of Sussex, and were adopted some time in the nineteenth century as formal breeds, and celebrated around the world.

This sheepish footnote to global history is simply an entrée to a bigger investment in the making of the modern world, by people who lived on the South Downs. The sheep breed begins at Glynde, where a master breeder called John Ellman turned promoter and began to market the Southdown lamb as an animal to be coveted, and the Southdown ram as the ideal animal to cover ewes. The Hanoverian King George III, sometimes known as Farmer George, a patron of scientific research and an encourager of the English Agricultural Revolution, thought highly enough of the venture to visit Ellman's farm.

By this time, England had lost its American colonies, and France had overthrown its royalty, established a republic and persecuted its aristocracy. Glynde is within walking distance of both Lewes and Firle, and men now forever linked with both places had roles in these events. We take our history as we find it, and we usually find it by accidental discovery: textbook history provides a skeletal narrative, but the past tends to come alive as we discover that history's actors were something more than names in the *dramatis personae* of revolution or civil war: they were people, fallible, tetchy and sometimes foolish; and we tend to learn a little about these incidental and sometimes innate weaknesses only when we explore places associated with them. Thomas Gage, son of a nobleman, was born at Firle Place in 1719. He joined the British army, fought in Flanders in the War of the Austrian Succession, which was nominally about the right of Maria-Theresa to head the Hapsburg empire; he also fought in Scotland, during the Second Jacobite Uprising, which was about the right of the Stuart family to the throne of England and Scotland. He then fought, alongside the Virginia-born soldier George Washington, in the Indian and French wars in North America, became military governor of Montreal, and after a series of adventures achieved the rank of commander-in-chief, North America, and finally governor of Massachusetts Bay Colony, too late for the Boston Tea Party but just in time for encounters with dissident settlers, among them Paul Revere and the Sons of Liberty. He was involved in the battles of Lexington and Concord, and called the Battle of Bunker Hill 'a dear bought victory, another such would have ruined us'. He was on his way back to England by 1775, and was buried in the family plot at Firle in 1787.

The other great figure of revolutionary America was Tom Paine, a tobacconist and excise officer at Lewes, who was encouraged to emigrate to the Philadelphia colony in 1774. The man who encouraged him was Benjamin Franklin, friend and companion to a number of British scientists, inventors and thinkers, and Paine arrived just in time to make an unstable situation even more explosive. A resourceful engineer and inventor, he was also a skilled campaigner, editor and pamphleteer with a remarkable gift for phrases that stick. 'The cause of America is in a great measure the cause of all mankind,' he began in *Common Sense* (1776), and then helped his readers distinguish between the entities of society and government: 'Society is produced by our wants, and government by our wickedness … The first encourages intercourse, the other creates distinctions. The first is a patron, the last a punisher.' He followed *Common Sense* with a series called *The Crisis*, and that too was marked by rhetoric that, once read, is remembered, and not just by the American patriots:

These are the times that try men's souls: The summer soldier and the sunshine patriot will, in this crisis, shrink from the service of their country; but he that stands it now, deserves the love and thanks of man and woman. Tyranny, like hell, is not easily conquered; yet we have this consolation with us, that the harder the conflict, the more glorious the triumph. What we obtain too cheap, we esteem too lightly: it is dearness only that gives every thing its value.

Populist firebrands who speak their mind with compelling phrases tend not to remain popular once revolutionary fervour has cooled a bit: Paine returned from revolutionary America, wrote *The Rights of Man* in 1791 and moved to France the following year. The country of his birth charged him with seditious libel, but the French elected him to the National Convention, even though he had not learned to speak French. Once again his gift for argument got him into trouble, this time with Robespierre: the following year he narrowly escaped the guillotine, and was released because the American ambassador claimed him as an American citizen.

Napoleon claimed to have slept with a copy of *The Rights of Man* under his pillow. Paine, when he saw the course of Napoleon's ambition, denounced him as 'the completest charlatan that ever existed'. He was not born in Sussex, nor did he die there, although every year Lewes makes a big thing about Paine from 4 July to 14 July, from Independence Day to Bastille Day. Paine was a Norfolk man, born at Thetford in 1737; he spent only six years in Lewes, and he died penniless and almost forgotten in 1809 in Greenwich Village, New York and was buried there: his bones were later exhumed by that other great English radical William Cobbett, and brought back to England. What happened to Paine's calcareous remains thereafter is not known, but there isn't much doubt about his legacy: President Barack Obama quoted Tom Paine on the day of his inauguration in 2009, and to mark the bicentenary of *The Rights of Man*, the brewers Harveys of Lewes made a pale ale called 'Tom Paine', so his health is celebrated in Sussex alehouses every day.

This chapter is not – how could it presume to be? – a history of Sussex, but a personal reflection on my bit of Sussex: the curious mix of bedrock, landform, foliage, commerce, habitation, tradition, tangential history, song, smell and remembered literature that goes to make up a line of an address. To first learn about Sir Thomas Gage, I had to visit Firle Place ('Home of the Gage family for over five hundred years'). But I cannot remember when I first heard about Tom Paine, because to learn his name, you just have to read the history of America, or France, or England: proof once again that the pen is mightier than the sword. The history of Sussex, like the history of any other shire, county, duchy, canton, commune, department, province or palatinate anywhere in the world, is rich in extraordinary figures thrown into sharp relief by the turns of history, and by the might of the pen. The connections with the county may be just as tangential as Tom Paine's. Percy Bysshe Shelley was born at Horsham; the young H.G. Wells educated himself in the library of Uppark, near Petersfield, where his mother was in domestic service. Both of them were remarkable figures, but nothing in their writings has a great deal to do with the county.

Rudyard Kipling, on the other hand, born in Bombay, now Mumbai, in 1865, married to an American, a world traveller, an apologist for the British Empire, first cousin to a prime minister, nephew by marriage to two distinguished British painters, the first Englishman to win the Nobel Prize for Literature, became a Sussex man by choice. He bought Bateman's at Burwash in 1902, and lived there till his death in 1936. A series of short stories and poems that date from the period began to establish East Sussex as 'Kipling Country'. This itself is a little odd: Sir Arthur Conan Doyle, begetter of Sherlock Holmes, Professor Challenger, Brigadier Gerard and the archers of the White Company, lived at Crowborough, East Sussex, for the last twenty-three years of his life. Who knew, or cared? Henry Rider Haggard, equally famous in his time, lived at St Leonards-on-Sea, East Sussex, for five years; he is, however, identified only with Africa, or Norfolk. Henry James, the American novelist, lived at Lamb House in Rye, East Sussex, for his last eighteen years, and Lamb House is now his monument, but I do not associate Henry James with the county in which I live. The difference is that Kipling made the Sussex landscape his own. The 1905 short story 'An Habitation Enforced' was first published in a US magazine. It tells the story of a couple of rich, rootless Americans who find themselves dispatched 'into that wilderness which is reached from an ash-barrel of a station called Charing Cross'. The only further geographical clue comes when the wife says, 'All this within a hundred miles of London.' Thereafter, all the proper names are fictitious, including the names of the properties they buy ('"Friars Pardon – Friars Pardon!" Sophie chanted rapturously, her dark grey eyes big with delight. "All the farms? Gale Anstey, Burnt House, Rocketts, the Home Farm and Griffons? Sure you've got them all?"').

I first read the story as a young teenager in suburban New Zealand, in a Kipling anthology that dates from 1946. I think I knew even then that he must be have been describing Sussex, and if I didn't know then, I certain knew by 1965, when I first walked at night several miles through farmland and woodland from Battle railway station to Netherfield, and to the house of my then freshly acquired wife's aunt, and felt that I already knew myself to be in Kipling country. I took a wrong turning, got lost, arrived in the wrong part of Netherfield, and entered an inn. Whatever was going on in the public bar stopped completely while I asked about the direction to Netherfield Cottage. The landlord, exercising the kind of authority displayed by those landlords who populate Dickens and Thomas Hardy, quizzed me at first suspiciously, then said, 'You want Mrs Pring's place!', turned to two young gypsum miners drinking at a table and said, 'Charles and Harry, you've had enough, you run this man down to Mrs Pring's.'

It was the first of an occasional series of encounters with Sussex dwellers that left me with the sense of intrusion into somebody else's fiction, or perhaps ghost story. That strange sense of a haunted landscape endures, and is evoked rather more deliberately in another Kipling story from the same volume. 'They' was first published in 1904, and this time it gives one firm geographical clue as the 'orchid-studded flats of the east gave way to the thyme, ilex and grey grass of the Downs', and another, a sentence or two later, finds the motorist-narrator at 'that precise hamlet which stands godmother to the capital of the United States'. Washington is a parish in the Horsham district of West Sussex.

But such clues are not necessary. The landscape of the stories seems to me palpably, inescapably Sussex. The puzzle is: how did he do it? The descriptions of woodland and open countryside are remarkably spare and deliberately imprecise. 'In the ungrazed pastures, swaths of dead stuff caught their feet ... But there stood the great woods on the slopes beyond – old, tall and brilliant, like unfaded tapestries against the walls of a ruined house,' runs a passage from 'An Habitation Enforced', and in 'They', a quick turn plunges the motorist 'into a green cutting brim-full of liquid sunshine, next into a gloomy tunnel where last year's dead leaves whispered and scuffled about my tyres'. I have read these stories again and again, and come to the despairing conclusion that, like the smile on the face of Leonardo's *Mona Lisa*, the Sussex location announces itself only to the reader's peripheral vision. The acute, precise sense of topographical identity, the *genius loci* or spirit of place, disappears as soon as you start a line-by-line search for it. La Gioconda's smile, too, becomes more enigmatic and humourless as you stare directly at the lines of her lips. Words are the props of a storyteller's conjuring trick, and the whole point of the conjuror's art is that you *cannot* see how he could have pulled the rabbit from the silk hat. Henry James described Kipling as 'the most complete man of genius (as distinct from fine intelligence) that I have ever known'.

On the evidence of some of Kipling's other short stories, I wouldn't swear to the fine intelligence either, but there isn't much doubt about genius, and especially the *genius loci*, and Kipling's obvious fascination with the spirit of place, that sharp but intangible sense that *here* felt very different from *there*, where one had thought one belonged, must have provided the basis for my own early obsession with travel, with seeing, feeling, hearing, smelling and tasting foreignness, with trying to understand other landscapes and other communities from the ground up.

I'm not sure when I first began to connect Kipling and Sussex (like everybody else who read the *Jungle Books*, I first identified him with India). The connection of Sussex with chalk, and chalk with Sussex, however, can be very precisely dated to a first year at secondary school, and an English literature textbook anthology that contained examples of the essay form. One of these essays introduced me to a prodigal and self-indulgent journalist and novelist for whom I still bear an immoderate and forgiving affection. The author was G.K. Chesterton, and part of his appeal remains that he was not simply a voice, and an avatar, a guide to a unique and gaudy version of London, in novels such as *The Napoleon of Notting Hill* and *The Man Who Was Thursday*; he was a member of a club of Edwardian entertainers, campaigners and arguers of profligate talent. These included the playwright and social critic George Bernard Shaw, the inventive, scientifically aware and politically engaged H.G. Wells, Chesterton himself, that French-born Sussex proselytiser and prolific prose-producer Hilaire Belloc, and the rather more self-effacing journalist and Russophile Maurice Baring; and as a member of this club of talkative table-thumpers, the search for more Chesterton led me to all of them. But in that one essay, written for the *Daily News*, and since preserved in a 1909 collection called *Tremendous Trifles*, Chesterton showed me what words and a simple idea, and a sense not just of humour but of comedy, could do.

The essay, with a title that echoes Huxley's, describes the author, who had attended the Slade art school, setting off from a Sussex village with a pile of brown paper and some coloured chalks; brown paper because he liked brown paper 'just as I liked the quality of brownness in October woods, or in beer, or in the peat-streams of the North'. He walks up into the Downs, 'those colossal contours that express the best quality of England', and crosses 'one swell of living turf after another', looking for a place to sit and draw. He doesn't want to draw from nature; he is going to draw 'devils and seraphim and blind old gods that men worshipped before the dawn of right'. It is at this point that he discovers a missing weapon in his artist's arsenal: on brown paper, he needs white. It is a confirmation that white is 'not a mere absence of colour; it is a shining and affirmative thing, as fierce as red, as definite as black'.

Chesterton was a newspaper essayist, writing regular columns to an agreed length: the survivors of this form know that they must begin with an anecdote, play with an idea, set up a paradox and produce a resolution that brings the anecdote neatly to a close. By the last paragraph, after a startlingly inventive hymn to the glories of whiteness, the writer racks his brain for expedients, and then begins to laugh: he is, of course, sitting on an immense warehouse of white chalk, a landscape made entirely from the stuff. He breaks a piece off the rock upon which he sits: it isn't as good as a shop chalk, but it gives the effect he wants, and brings him to a fundamental understanding. It is the one with which we began this chapter, one explored by that other great essayist Thomas Henry Huxley, and one that established within this then very young New Zealander the first intense awareness not just of a particular landscape ten thousand miles away, one that was home to little villages, county towns and itinerant artists and therefore an identifying address; but of the underpinning to any landscape, of the bedrock on which a local civilisation must be built. 'And I stood there in a trance of pleasure,' said Chesterton, 'realising that this Southern England is not only a grand peninsula, and a tradition and a civilisation; it is something even more admirable. It is a piece of chalk.'

FOUR

The Country

England, Their England

England is what Shakespeare called 'a local habitation and a name', but it is not in the strict sense a postal address. By 1840, when Rowland Hill created the uniform penny post, and reformed the nationwide delivery service, England was not a sovereign kingdom. For most of my life, England was part of an administrative entity called 'England and Wales', which was itself just one component of the United Kingdom of Great Britain and Northern Ireland. Furthermore, for at least the first half of my life I observed something complacently self-effacing, or even smugly deprecating, in English attitudes to England. People might stress a Cornish heritage, or Cumbrian traditions; they might make patriotic distinctions between identities as a Man of Kent, or a Kentishman; they might keep alive memories of the Wars of the Roses in a Yorkshire–Lancashire rivalry; but they were often quite embarrassed about the notion of Englishness itself, and would volunteer that the Scots had better engineering skills and grander landscapes; the Irish had more charm and jollier pubs; the Welsh had more song and certainly more poetry.

England existed as a sporting identity – cricket, rugby, association football and so on – and there were certainly always vociferous England supporters, but even the most vociferous would admit to surprise when England won anything. Welsh nationalism was never far from the news pages. Scottish identity was often a source of fierce and even prickly pride. The English, when they talked about Englishness at all, tended to present themselves as degenerate, on a downhill slide, of no particular consequence, coloration or identity. English writers who visited New Zealand would sustain the fantasy that the New Zealanders of that time were more English than the English, a remark which only reveals its fatuity when you transpose it to proclaim that the English are less English than the New Zealanders.

There was, and still is, a national day – St George's Day, also widely accepted as Shakespeare's birthday – but 23 April was marked only by querulous letters in certain newspapers lamenting the absence of celebration of St George and Englishness. In the ensuing correspondence someone would point out that St George was a Roman soldier from Cappadocia, now a region of Turkey, who died as a Christian martyr in Palestine, who almost certainly never set foot in England, who did his entirely mythical dragon-slaying in Libya, and who is also the patron saint of Aragon and Catalonia, and other places once at war with England. Even now, the Royal Society of St George, which describes itself as the 'premier patriotic society of England', says that the celebration of St George's Day in England is 'currently fairly low key'.

But this sustained demonstration of humility, this collective, well-bred wrinkling of the nose at words such as 'English patriotism', fooled nobody except possibly the English. It seemed at the time to mark a collective self-satisfaction, so deep and so secure that most people felt no need to express it, and if they did were more likely to apologise, quite diffidently, for their sense of national superiority.

This self-assurance, at the start of the 1960s, was a structure with almost no visible means of support. The phrase 'good English food' was used a lot, but it was an oxymoron; the urban

English, who at the time vocally suspected foreigners and foreign food of being 'dirty', were a people who lived in grimy, draughty, damp and usually freezing houses, often without bathrooms; who wore the same shirts for a week (changing the collar from time to time) and who inhaled with each breath a gruesome mixture of soot, sulphur, pollen, cigarette tobacco and the aroma of stale frying fat. They looked nothing like the people that so impressed Pope Gregory the Great, who, when he asked about some blue-eyed, fair-haired slaves from Kent, and was told that they were Angles, is supposed to have quipped, '*Non Angli, sed angeli*: not Angles but angels,' and in AD 596 dispatched St Augustine of Canterbury to convert the heathen English to Christianity.

The English of 1961 were neither fair-haired nor angels, but they seemed to have a certain saintly patience. They endured overcrowded public transport, perpetuated preposterous class attitudes and deferred to a dismissive bureaucracy, and while they defended the possession of an empire they had, most of them, never seen, and were sullenly struggling to retain, against all moral and pragmatic logic, they also prided themselves on their insularity. They enjoyed the illusion though not the reality of great maritime resources, but were quite happy in subsequent decades to close the shipyards of the Clyde and the north-east, to hand over ownership of the great shipping fleets to other nations, and to give up within a single generation a national tradition of seafaring. They sometimes rather patronisingly referred to Australians and New Zealanders as 'colonials' and sometimes confided wistfully of emigration plans long abandoned; they complained sullenly of the 'Americanisation' of British culture but queued to see Hollywood movies in preference to anything from Pinewood; and their musical answers to the phenomenal Elvis Presley and Bill Haley were the somewhat lower-voltage Cliff Richard and Frankie Vaughan. Their classical music was performed by people with Russian, German or Italian names, the films most hotly discussed in the cerebral weeklies and Sunday papers were almost invariably made in France, Italy or Bengal, and for a short period the most successful single playwright in the West End was the very French Jean Anouilh, with productions in five theatres at the same time. London's buildings were dirty, its cafés unwelcoming, with bitter coffee and nauseating provender, and there was no place on Earth I would rather have been.

That the England of 3 February 1961 – the date of my disembarkation at Southampton – was cold, grubby, war-damaged and depressed was of no consequence. This depression masked another England: one that existed almost entirely in spoken and written words, but that had occupied a place in my mind as palpably as any physical address. Some of this may safely be blamed on Shakespeare and Dickens: we had all learned by heart at school the words of John of Gaunt's death-bed speech, the one about a precious stone set in a silver sea. We had all of us already been to Bleak House and Dotheboys Hall and the Essex marshes where Pip encountered Magwitch. We knew about Arthur and Merlin and the Round Table: it was in that year that an American presidency identified itself openly with Camelot. We could all recite passages or whole poems from Wordsworth and Tennyson, Shelley, Chesterton and Walter de la Mare. We listened eagerly – television did not arrive in New Zealand until after I had left – to John Gielgud in a dazzling BBC radio production of Christopher Fry's *The Lady's Not for Burning*; to the awful jokes of Frank Muir and Denis Norden's *Take It From Here*; to the anarchic invention of Spike Milligan and the gang in *The Goon Show*; to the urbane drollery of panel games such as *Twenty Questions* and *My Word*; to recordings of Richard Burton reading *The Rime of the Ancient Mariner*.

But England was more than just the home of Prince Hal and Falstaff, or the headquarters of the Pickwick Club, or the place where Thomas Mendip wanted to be hanged, or 23 Railway Cuttings, East Cheam, home of Anthony Aloysius Hancock, in *Hancock's Half-hour*. It was the seat, the palace, the powerhouse of an empire of words: not just home to the English, but to English itself, the physical address that gave me the words for address, road, town, county, country, kingdom and so on. I had – I can date this to my fifteenth year – a very clear understanding of the difference between British English and American English, because I loved the New York stories of Damon Runyon, the wonderful language of Herman Melville's *Moby-Dick*, and the invention of Mark Twain. I also enjoyed reading the *Saturday Evening Post*, a journal that seemed not to acknowledge the existence of anywhere except rural and small-town America; and two other magazines that did monstrous but mesmerising things to the mother tongue. One of them was Henry Luce's *Time* magazine, which developed a literary style distinctive enough to be routinely parodied ('Backward ran the sentences, till reels the mind ...' went one sardonic anastrophe in the Lucian style). But it also delighted in verbal quickness of wit. In 1960 it described an episode in which the bellicose Frank Sinatra 'concerned as ever to prove that he is no pip-squeak, pip-squawked: "Can you fight? You'd better be able to."' The other was the US *National Geographic*, never thrown away, so that on a rainy Sunday, old copies provided fresh marvels. One showed a picture of a placard that said 'Abandon hope all you who enter here' over a dugout on a wartime Pacific island, and then extended my understanding of the possibilities of language with the caption 'Laughs Aimed at Themselves Keep Marines' Morale High in Oft-bombed Guadalcanal'.

One of the first books to make me aware of the potency of words, rather than the stories they told, was indisputably

English, and it dates from the era in which American and English verbal traditions had yet to diverge. It was John Bunyan's *The Pilgrim's Progress*. I read it completely uncritically; I read it long before I learned about words such as 'allegory', and I remember being made anxious by the geographical menace of Doubting Castle, the Slough of Despond and Vanity Fair, all airy nothings that meet the prescription for name and address in *Midsummer Night's Dream*:

> And as imagination bodies forth
> The forms of things unknown, the poet's pen
> Turns them to shapes and gives to airy nothing
> A local habitation and a name.

But I can also remember being thrilled – a prickling of the skin, a sense of fierce excitement – at the simple succession of words that took Mr Valiant-for-Truth to his final destination:

> When the day that he must go hence was come, many accompanied him to the Riverside, into which as he went he said, *Death, where is thy Sting?* And as he went down deeper he said, *Grave, where is thy Victory?* So he passed over, and all the Trumpets sounded for him on the other side.

It was in Bunyan, too, that I must first have noticed the potency of that word address: 'He addressed himself to go over the River.' The English word address has a long history and a huge number of meanings, but all of them are contained within its etymology. The word arrived with the Normans, but began somewhere long ago in Roman Europe: it is a composite of the Latin prefix *ad*, meaning to, or towards; and *directum*, straight, or direct. This ancient coinage already contains within it the possibilities of direction, readiness, attitude, diction, discourse, inscription and attire. According to the *Oxford English Dictionary*, it appeared in various spellings more than six hundred years ago as an English verb: to straighten, to erect, to set up, and also to put to rights, to arrange, to put in order. 'A Parlament being call'd,' wrote John Milton in 1670, 'to addres many things.'

Only seven years later, in *Paradise Lost*, Milton uses the same word in the sense of direction: 'So spake the enemy of mankind and towards Eve Addressed his way.' But by then it has already come to mean words in a letter, intended for a specific person. 'For the Advice is addressed to the Sovereign only,' says Thomas Hobbes in 1651, in *Leviathan*. At the same time, it also meant direct speech to someone: prayers could be addressed to God, speeches to the House of Lords. By 1712 it had become the name and place to which a letter might be directed. By 1880, the meaning had widened to become not just the person, and where he or she might be found, but the place of residence: an address that existed independently of its tenant or proprietor. Similarly, every line of a modern address contains the same magical, shifting history.

I live on a road, in a town, in a county, which is a defined fragment of a country. Both the road and the town have proper names that no longer mean what they once meant. But the generic names, too, have histories that tell me about the past. For most of my life I have assumed, from the sense in which such words are normally used, that a road was different from a street: that a road went somewhere, whereas a street was a thoroughfare that existed to serve the houses on each side of it. A street was a local convenience for townspeople; a road was their way out of town. We mark this distinction in our truisms, commonplaces and our songs: all roads lead to Rome, hit the road Jack, the road to hell, the Old Kent Road; on the sunny side of the street, the street where you live, right up my street, Coronation Street.

This distinction is, I learn, not supported by the history of our words. The *Oxford English Dictionary*'s definition of a road as a line of communication between different places, 'usually one wide enough to admit of the passage of vehicles as well as of horses or travellers on foot', is only its fourth meaning. In AD 888, in his translation of the Latin philosopher Boethius, King Alfred used the word 'rade' to mean a variant of riding on horseback. The same word became 'raid': an attack by mounted men; it also came to describe the riding of ships at sea; and the sheltered roads in which ships lie at anchor, offshore. The meaning we now use is quite recent: the earliest references to roads as highways, or addresses of domicile, pop up in the Elizabethan playwrights. 'I think this be the most villanous house in all London road for fleas: I am stung like a tench,' complains a carrier in Shakespeare's *Henry IV Part 1* in 1596. The sense of a road as a common highway along which people would ride, or walk, emerged over the next two hundred years.

Street, on the other hand, really is an old word. It occurs in various forms in Old English – the poem Beowulf contains the word '*straet*' – old Frisian and early Danish, old German and ancient Irish, but all these variants derive from the Latin word *stratum*. To the Romans who once built roads all over Europe, for the convenience of marching armies and galloping messengers and rolling wagons laden with wheat and oil, *via strata* meant paved road. We still use the Roman meaning when we identify certain Roman roads: Watling Street, Ermine Street. The meaning of street as a broad highway and road as a bridleway or horseman's route lingered for centuries: in 1881 the Victorian novelist R.D. Blackmore wrote of 'a place where the street narrowed into a road'.

We are on a journey to the idea of address, but this journey takes us into etymology's byways. We cannot understand a word properly until we learn its history, and when we do, it becomes more precious to us. Words are the foundation of wealth and potency, the coin with which a community cooperates, unites and exercises control over its environment. They are the currency of ideas, but these ideas acquire precision only if the words that express them are understood in the same way by the speaker and the listener. So the idea of a lexicon must have existed, in some unwritten form, long before Dr Johnson laid the foundations of lexicography with his dictionary, and the meanings of words must have been handed from generation to generation by two routes. One of them would have been formal pedagogy: a pronouncement or a dictum from an appointed tutor or storyteller. The other would have been the more natural method: we listen to the sense of the sentence, and unpick the meaning of the word from its spoken context, and perhaps disinter a previous use from a written document. Inevitably, spellings mutate, emphases shift and meanings change. But there is something thrilling in finding a connection that extends across two thousand years of history: that the mundane and almost redundant word street (Americans frequently abandon it, and say things like 'At the corner of Hollywood and Vine') should have a direct connection to the eagles and the standards of the Roman legions who rode or strode along highways that ran straight across the landscape, on foundations of gravel, and paving, and ambition, from posthouse to camp, and from redoubt to town.

The word town, too, tells a story of changing human settlement. A town is a concentration of buildings used for worship, civic and judicial administration, business, trade, entertainment and domicile. These buildings are divided by, served by, identified by roads, streets and avenues and other terminology associated with address and thoroughfare, including boulevards, crescents, squares, courts, fields, alleys, hills, lanes, yards, terraces, crofts, mews and gardens. In common English usage, according to the *Oxford English Dictionary*, the word town denotes 'an inhabited place larger and more regularly built than a village and having more complete and independent local government'. This description implies a formal boundary, where the local government jurisdiction ends, and the usual way to leave town is by road. This common and hardly to be questioned usage, however, is a relatively recent one. Long before King Alfred, the first meaning of the Old English word town or *tun* or *tune* was simply an enclosure: a field, yard or court; town came to mean the property around a farmhouse or manor; then a group of buildings around a farmstead; and then a hamlet or village, before it finally achieved its present meaning. Some time in the seventeenth century, a walled spring in Kent, near the East Sussex border, began to grow into Royal Tunbridge Wells. But we salute the Old English garden or orchard or sheep pen not just in the word town but in the names of hundreds of towns: Wellington, Brighton, Shepperton, Taunton, Tonbridge and so on. In the same way, the Norse word *by* meant farm or settlement, and came to mean town, so the region raided and then colonised by Vikings in later centuries is marked by towns that end with a different suffix: Whitby, Rugby, Derby and so on.

The word county came over with William the Conqueror, who awarded domains and territories to his counts or *contes*, but the word was already ancient by then: it too derives from Latin. A *comitatus* was the retinue of companions with whom a warrior might break bread. A count would have been a warlord, a provider of men, horses and armaments, who gambled on a king's military success and hoped to gain land and wealth from the wager. A king had every interest in rewarding his dukes, earls and counts, because their hold on their territory, and the taxes they levied, in turn helped him retain the throne he had seized. It isn't hard to see how the lands held by a count gradually became administrative units, especially when some of these had already become administrative territories or shires under the English kings. A county was a seat of power, a register of the nobility's lands, a jurisdiction, a place where an aristocratic word could be law. The word still carries within it implications of gentry as well as of autonomous administrative authority that might either support or defy a king.

The word England, too, is an imported one: it is Old English, but was shipped across the North Sea from what is now Germany: the word Angelond is Old Frisian. The Venerable Bede, author of the *Ecclesiastical History of the English People*, in 731, called the natives *gens Anglorum*, the race of Angles. The word English, in its various spellings, once referred to the blend of dialects spoken by the Angles and Saxons who invaded and colonised Roman Britain, and was to be distinguished from the language of the educated world, which was Latin. English-speaking people who use words for a living still make a similar distinction: English words tend to be shorter, more direct, more vivid than those derived from Latin. You can be effulgent and felicitous, or you can be bright and happy. The meanings are more or less the same: the impact is not, and the happiest and the brightest way to use the language is to select the English word where possible. On the other hand, if you are going to coin, invent or fabricate a new word to describe a newly invented, discovered, or imagined thing, the languages of Rome and Athens are traditionally pressed into service: telephone, telescope, teleportation. But even these coinages are unconsciously exchanged for casual English: 'Give me a call,' we say. 'Look through the glass.' And far in the future, when Captain Kirk of the USS *Enterprise* wishes to teleport himself from the danger zone, he resorts to very old English: 'Beam me up, Scottie!'

Words are instruments, they are tools that, in their different ways, are as effective as any sharp edge or volatile chemical. They are, like coins, items of great value, but they represent a currency that, well spent, returns ever greater riches. Coins are passed from hand to hand: detectors of sufficient sensitivity could, in theory though hardly in practice, trace their history by identifying local dust and biological material trapped in the patina of the metal after each handling, and separately identify the cycles of purchaser, shopkeeper, employer and banker, all the way back to the mint. Words, too, come loaded with their own past: they are their own fossils, and their origins contain clues to the ways in which they will subsequently be used. All the words that comprise an address have preserved histories: bungalow and veranda were borrowed from Hindi, and are casual reminders of two centuries of British pillage and purchase on the Indian subcontinent. House comes from the Old English, and shares its ancestry with similar words in Old Frisian and Old High German. Villa on the other hand was a Roman word for a country house that was the origin of *ville* in France and village on both sides of the English Channel, and was somehow revived in its original meaning to describe suburban or country homes in the English eighteenth century. But because a country property might be cultivated by a subject peasant or a feudal serf, the word villa also engendered the villain: once a base or low fellow, and then by association a scoundrel. Manor, mansion, manse and maisonette all spring from the Latin word *maneo*, to stay, tarry or dwell: manor, however, already existed in Old English, while mansion and maisonette arrived from France. Manse, a measure of land or dwelling for an ecclesiastical family, may have been lifted directly from medieval Latin, which of course was the language of the Church everywhere in Europe. A cottage comes from a medieval Latin word that described the home and garden of a feudal tenant, cotter or cottar; but the word cot, for cottage, and cote, as in dovecote, already existed in Old English, so at some point an Anglo-Saxon word was adopted by French feudal lords, Latinised, and handed back to English householders.

The words we use now, like the things we live in now, are just the visible ends of long threads through the tapestry of history. Sometimes the threads are not so long: apartment and avenue are relatively recent borrowings from the French. Avenue, in the sense of approach or access-way, usually lined with trees, pops up in English usage only in the seventeenth century, when ambitious property owners became concerned with the appearance of property; the diarist John Evelyn describes walking up an 'ungraceful' avenue in 1654, but he has already in 1641 recorded moving into new lodgings, 'a very handsome apartment just over against the Hall-court'. Words such as boulevard, in the sense of a wide street or promenade, first appeared in Paris in the eighteenth century, but actually date from a Teutonic word, *bollwerk* or bulwark, because the first boulevards were the flat bits of the ramparts of a fortified town. A residential road, street or close of the kind called cul-de-sac is a thoroughfare open only at one end: the phrase means 'bottom of the bag' in French, but people in France call such a street *une impasse*, or *une voie sans issue*. There is a kind of residential cul-de-sac called a close. The word first appears in the meaning of an enclosed space in 1297: its origins are French, but the word derives from the Latin *claudere*, to close or shut something. The word cloister springs from exactly the same source, and appears in English at around the same time, but Old English already knew about cloisters: it already had monasteries, Latin and *clausters* to which monks could retreat to extend their scholarship.

The words we use are part of a huge, teeming, shared history; their meanings shift with the decades, because their use is tied to particular places and times: they leave a trail through written history. Words for house, and street, and road and cottage, county and country remind us that the land we live in is not, in any sensible way, ours by right, but by might. Julius Caesar landed somewhere near what is now Walmer, or Deal, in August

55 BC, and established a beachhead; he wasn't the first invader, because a tribe of Belgae had crossed from Gaul and already established themselves in Celtic Britain more than a hundred years earlier. But very few Celtic words live on in the names for our highways and byways, our hearths and homes. Highway, byway, hearth and home are all words of Teutonic origin, and the word way may be much, much older, and its meaning, says the *OED*, has been 'to some extent influenced' by the Latin *via* and the French *voie*. So even if you knew no history, you could stare at the words we have and build up a picture of a country with a small population, dispersed into smaller settlements, colonised, dominated and changed forever by Latin-speaking invaders from Italy, who introduced the English elm tree, and hot baths, and who built a barrier against the northern peoples; who maintained the peace and established a series of permanent camps or *castra*, that are now acknowledged in place names with the suffix *chester* or *caster*, and who then finally left: withdrew to Rome to defend the eternal city against the new menace.

The Romans were replaced by systematic marauders: fighting seamen and farmers from the other side of the North Sea, who certainly spoke different dialects but who somehow forged an identity that over a few centuries became Angle-land or England, and a language strong enough, and confident enough, to withstand, survive and finally absorb invasions from Viking Scandinavia in the north-east, and from Norman France in the south. One Germanic tribe called the East Angles settled in what is now East Anglia. My county of Sussex was settled by South Saxons. Cities that already existed continued to exist and grow. The Roman settlement of Eboracum became Jorvik and then York, Londinium became London, and so on. After 1066, the invasions continued, but peacefully, and in small increments: increasingly diplomats, merchants, skilled craftsmen, scholars and religious and political refugees came and went, or came and stayed, wrote letters home, and compiled descriptions, memoirs and diaries that recorded where they lived, whom they visited, how they travelled, the roads they took, the apartments they rented, the homes they built. Long before I came to England, I had a geography of unconscious addresses: Dr Johnson lived near the Cheshire Cheese off Fleet Street; Samuel Pepys lived in Seething Lane; Dickens at Gad's Hill; Isaac Newton at Woolsthorpe Manor; Darwin at Down House, Downe, Kent; Rudyard Kipling in Villiers Street, or Rottingdean and then at Burwash in Sussex, as well, of course, as at Naulakha near Brattleboro in Vermont, and the *Civil and Military Gazette* offices in Lahore. But just as potent were the addresses that existed principally as fiction: the Hundred-Acre Wood of *Winnie-the-Pooh*; the great house of Chesney Wold, Krook's rag and bone shop near Chancery Lane and Mr Jarndyce's Bleak House from the Dickens novel of the same name; the boarding house of Mrs Todgers in *Martin Chuzzlewit*; the rooms rented by Sherlock Holmes and Dr Watson in Baker Street; and Enmore Park, West Kensington, the home of Conan Doyle's other great creation Professor Challenger.

Some writers seem inseparable from place. The novels of Dickens sketch a London geography so palpable and so coherent that you might use them as a crude, remembered map: I worked for decades in Clerkenwell, and never forgot that I was at the confluence of *Barnaby Rudge* and *Little Dorrit*: it helped, of course, that habitations such as the Charter House were remembered in Charterhouse Square, and even though the Plornishes had never existed, Bleeding Heart Yard certainly looked like the home they might have found there. I could even offer an explanation of why I could not personally find Mrs Todgers's boarding house. That was because it was in a labyrinth: everybody knew that. Dickens had already provided the excuse:

Instances were known of people who, being asked to dine at Todgers's, had travelled round and round for a weary time, with its very chimney-pots in view; and finding it, at last, impossible of attainment, had gone home again with a gentle melancholy on their spirits, tranquil and uncomplaining. Nobody had ever found Todgers's on a verbal direction, though given within a few minutes' walk of it.

Nobody ever established his London as intimately as Dickens did. The London of Sherlock Holmes is more shadowy, but any detective who cites Bradshaw's railway guide as a forensic tool is not going to play fast and loose with physical geography. The works of Stevenson play more imprecisely with landscape: I could not direct you to the Admiral Benbow, Black Hill Cove, in *Treasure Island*, but I can still hear Blind Pew tapping his way along the path towards it.

Some obsessive readers want to identify the people who inspired the fictions that stalk forever through great books, and it makes a good game. I was delighted to learn, from William Amos's 1986 book *The Originals: Who's Really Who in Fiction*, that the original model for Richmal Crompton's grubby, mischievous outlaw Just William served in the Royal Air Force in Iceland with the original model for W.E. Johns's indestructible flying ace Biggles. But I never stopped to think: what larks these two boyhood heroes might have had together! I knew – everybody knows – that William Brown and James Bigglesworth are not the same as Crompton's brother John Lamburn and Air Commodore Cecil George Wigglesworth. The first two are immortal; the other two were not. But great books impose a second and compelling geography that overlays the physical topography and temporal structures of the mundane world. Marcel Proust's Albertine remains a beautiful young woman, no matter how often I am reminded of Proust's real appetites. But

Proust's Combray haunts the little French town he recalled so powerfully that in 1971, on the centenary of Proust's birth, the commune of Illiers surrendered and changed its name to Illiers-Combray. Why should we think that odd, or enterprising, or exploitative? Many more people know Proust's Combray than have ever been to Illiers, and know it far more intimately.

At least Proust's Combray really was, in part, a memory of Illiers. Critics, travel writers and historians talk of 'Kipling's India' as if it had substance, and could be separable from the real India, but someone who has read *Kim*, or the *Plain Tales from the Hills*, or the *Jungle Books* already knows an India, and tends to see, beyond Delhi airport, something already recognisable, already indexed in the gazetteer of memory. There may be less justification for such observations: Kipling's India was a composition that disappeared in reality a hundred years ago, and legally and administratively more than sixty years ago, when some of it became Pakistan. But it remains difficult to escape the sense that we 'know' Kipling's India, and see it immediately in the streets of Delhi. Good or merely exciting movies have the same effect: English people who step off an aeroplane for the first time in Los Angeles or New York are surprised by how strongly they feel that they 'recognise' a place they have never seen before. But many Hollywood movies are filmed on location, or on soundstages constructed with an eye to verisimilitude. New York really does look a lot like a movie.

The more astonishing realisation is that a few sentences, a string of black characters on a white surface, can and often do the have the same effect. Sometimes the literary creations are all that remain, and we deliberately reshape a topography and give it coordinates to match these fictions. Homer's Ilium and Agamemnon's Mycenae may, for all I know, have never existed: the people, the places, the bloody stories of rapine, revenge and destruction may be just that, stories, the ancient Greek equivalent of *Star Wars*, or *Lethal Weapon II*. But we make pilgrimages, now, to a city that somebody has unearthed and identified as Troy; and to a Peloponnese fortress that encloses the treasury of Atreus and the tomb of Clytemnestra, and evoke solemnly for ourselves amid the silent, sunbaked stones a set of episodes that seem as powerful now as they must have done two or three thousand years ago.

Words construct a reality for us, a phrase, a sentence, a paragraph at a time, a character and an episode at a time. England existed for me, in a mosaic of literary landscapes, long before I first saw the place, and when I did, I could not help seeing it as if, somehow, I had seen it before. London in 1961 was still afflicted with 'pea-souper' fogs; it was indescribably dirty, the limestone and brick blackened by soot from 150 years of smoky chimneys; old men in grubby coats played hurdy-gurdy machines and sold freshly roasted chestnuts on the street; slum-clearance programmes had already begun, but huge tracts of impoverished Victorian housing survived both in west London and the East End, and Tower Bridge was regularly raised and lowered to let shipping through from the Pool of London. It looked like the London of Margery Allingham, and G. K. Chesterton, and Charles Dickens.

On my first visit to meet my mother's family in Derbyshire, I was driven past a steelworks: the words 'dark, satanic mills' forced themselves upon my consciousness. It looked dark, it looked satanic; it seemed to smear the bright skies of the Derbyshire hills. The little, huddled brick towns and villages of the same countryside evoked images from D.H. Lawrence's *Sons and Lovers*, and the first time I walked through Paddington – its skyline pierced by huge cranes erecting high blocks of flats – I found myself murmuring those strange lines from Blake's epic 'Jerusalem':

The Country

> What are those golden builders doing
> Near mournful, ever-weeping Paddington?

The reality is that fiction is another country, not so easily separated from the one we think we live in, and I had been to England long before an Italian passenger ship carried me there, and deposited me at Southampton, with a weekend bag, a suitcase and two Penguin paperback translations: one of them *Don Quixote*, and the other Dostoevsky's *The Idiot*. I didn't think to carry any Dickens with me. I was going to Dickens country: I would find him and his streets all around me.

FIVE

The Nation

How to Unite a Kingdom

The act of union that first joined Scotland to England and Wales was haphazard, unpremeditated and witnessed by no one. Its impact was tremendous, its consequences colossal, its long-term outcome unpredictable, and its process epoch-making. It brought with it profound lessons for humankind, although these would not be learned for another four hundred million years. And much of it happened underwater.

Wind back the clock five hundred million years to a very different world. You wouldn't know the place. Nothing that you can see above the waves looks anything like any island, archipelago or continent that exists now, and quite a bit of the islands- and continents-to-be are below the waves. Think of islands and continents as mere bits of crustal mosaic, to be played with, to be stained, planed, chipped and moved around, but ever so slowly, to be bashed against each other and crumpled into one big continent and then broken up again by some mindless, heedless force deep beneath the crust. There is a bit that eventually will become England and Wales and part of Newfoundland, but it is deep in the southern hemisphere. This terrain has, for scholastic convenience, been given a name: Avalonia. It takes this name not from the Vale of Avalon, in Somerset, but from the Avalon Peninsula in Newfoundland, where some of the ancient rocks of Avalonia are exposed. The

Avalon peninsula, of course, was named for Avalon in Somerset, by English colonists who landed there in 1623.

But that is to anticipate. At this time there is no England, no Scotland, no Europe, no America. The fragments of sediment, rock and volcanic ash that are to become Scotland, too, are south of the Equator, and in what are now the tropics: this landscape has been called Laurentia, and some of it survives today as the St Lawrence Seaway, the Mississippi Basin, the prairie, Hudson Bay, Newfoundland, Nova Scotia and Greenland. What separates these landmasses are four thousand kilometres or more of ocean. This ocean too has a name, even though it no longer exists. It is called Iapetus, and it is closing. Old ocean crust is being destroyed, as – because it is colder and therefore more dense than other crustal rock – it slides down below the landmasses of Laurentia and Avalonia. It can go nowhere else, because Laurentia and Avalonia are riding towards each other on tectonic plates, moving at the rate of a few centimetres a year. These two landmasses are being shaped and altered as they move. The act of subduction, the diving down of ocean floor below a continental crust, to disappear back into the hot mantle, is a slow but also a violent one, accompanied by earthquake and volcanic eruption and stresses that cause the continental crust to crumple, and pile up on itself. Such things are happening right now: some of the crust of the Pacific Ocean is disappearing below modern Peru and Chile, which is why the Andes are still growing. Japan is a volcanic reaction to similar submarine violence on the other side of the same ocean.

So the rocky crusts of Laurentia and Avalonia are being altered by processes that were happening a billion years ago, and are still happening: because they are still happening, geologists can examine the process, record the consequences and use the experience to reconstruct things that might have happened five hundred million years ago. The bits of Laurentia and Avalonia that are above sea level are under continuous assault from wind, rain, glaciation and summer heat. Rocks that swell and shrink with changes in temperature are more likely to shatter into ever smaller fragments as the seasons change. Rocks exposed to rain will begin to dissolve. Some of the minerals in the rocks will flow into the sea as dissolved salts. Some will remain stubbornly resistant but will become smaller and smaller as they are bashed, smashed and abraded by their movement against each other, as they are ground by glaciers, swept by torrents and washed by advancing waves: they will eventually become mud, or sand, or shale. Whatever happens, they will erode. Time and tide are part of the planetary recycling process: forever turning mountains into molehills, and molehills into coastal sands and estuarine mud.

But while rock above the waves is destined for destruction, a different creative process is at work beneath the sea. Rivers that run off the land deposit their burdens of silt onto the seabed, building up a new platform for some yet-to-emerge landscape. Little creatures, and some not so little, are living in the oceans, feeding on plants, plankton and each other; forming shells and skeletons of hard carbonate material, and then dying, to bequeath their bodies to the submarine shelf. A lot depends on depth, temperature, available light and the supply of nutrients in the oceans, but this process slowly builds the shales, sandstones, corals, fossil limestone and chalk known to geologists as sedimentary rock. It starts off as soft, runny, mobile sludge, but give it time and heat and some high pressure from the sediments on top and it turns to rock.

There is a second process that makes new continental crust, and this too is amending the shape of both Laurentia and Avalonia. Volcanoes slurp magma, or explode and shower cubic kilometres of blazing ash everywhere, and this, too, gradually builds up into the fabric of an island, or a continent, to leave a signature of sudden, disruptive intrusion among the slowly growing beds of limestone and shale. Wait for a few million years for the climate to change. If it gets warmer, the ice melts and the oceans become more productive. If it gets colder, glaciers pile up on the sand, the sea level falls and new bits of land are exposed again to the wind and rain. Such things happened, again and again, to Laurentia and Avalonia, as they shifted imperceptibly but inexorably to a date with history: our history, not theirs.

This process, of course, involved more than just two little bits of what was to become western Canada and northern Europe. Avalonia is a fragment already detached from a massive, Antarctic supercontinent called Gondwanaland. It isn't the only detachment: Australia, Africa and South America have yet to part and go their separate ways. India will be plucked away and hurled northwards, to smash into a landmass with a force so colossal that it will create the Himalayas, and lift the Tibetan plateau towards the skies. The landmass waiting for violent intimacy with India is already free and riding on the oceanic crust of Iapetus. One part will become Siberia. Another is a province called Baltica. They were once huddled together in a supercontinent, and they will meet again. This is the history of continental fragments: to ride about, helplessly, on the tectonic plates that carry them together, again and again, and then separate them in different formations. It happened before, and it will go on happening, because the Earth is a living planet: it shapes, consumes, metamorphoses, destroys, recreates and recycles its components of earth, air and water.

It has made Pangaea, and Gondwana, and Laurentia, and Avalonia: it is about to make something that will four hundred million years later become recognisable as the United Kingdom. This process has already begun, piecemeal: across a narrowing sea, southern Scotland and northern England are approaching each other. There are volcanoes off the coast of England: one of them will become Snowdon, and granite has begun to seep upwards through the ash and the muddy shales to become the core of what is now Cadair Idris. The landmass of Scotland has begun to crumple into what will become the Grampians. Very deep sea has slowly accreted mud, silt and sandstone thousands of metres in thickness which will eventually pile up and heave above the surface to become the Lake District, along with thousands of metres of basalt and other volcanic rocks. Some of this violence is the result of a collision, but not yet with Scotland: Avalonia has already bumped into Baltica. Just as two cars in collision register the event with mutual crumpling, so can two landmasses. The speeds are dramatically different, but the cumulative momentum of a continent-sized traffic accident is awesome. The geological period during which these events take place is the Ordovician. It lasts for fifty million years. It is named after a tribe that once occupied Wales, which is where some of this story was pieced together.

The next part of the story – it continues through epochs, and crosses geological periods – occurs in the Silurian, a name also taken from a tribe of Britons in Wales. This is another twenty-million-year stage in the making of modern Britain: Avalonia is now in tropical waters, there are volcanoes in Pembrokeshire, and Wenlock and Ludlow are submerged, sandy and strewn with coral reefs. By the end of this time – how lightly we speak of twenty million years, in the context of the making of our present address! – the ocean of Iapetus has completely gone: the space once occupied by sea is filled with land. In a hundred million years, four thousand kilometres of ocean have been closed and two lumps of land have begun to pile head-on into each other, and the suture between the two is marked by the raising of mountains of Himalayan stature, and the rocks at the collision zone will be melted, deformed and metamorphosed into slates, schists and gneisses that will survive as testimony to the violence of the event.

The new mountain zone will bear no resemblance to modern Scotland: not with mountains that soar eight thousand metres into the clouds. But rain, snow, ice and wind will immediately begin to plane, chisel, sand down and all but demolish this new orogenous zone. There are still three hundred million years to go, and a new geological era is about to begin. It will be called the Devonian. It will be characterised by baking-hot equatorial skies, long periods of aridity and the formation of dunes of red sandstone of the kind first identified in the rocks of Devon. The resulting land formations will look almost nothing like the British Isles: the topography will be different, the coastline will be different, and of course the location is still equatorial. The Atlantic Ocean does not exist. What will become North America is still stuck to Scotland, awaiting subsequent tectonic surgery that will form a new ocean and make new continents of old landmasses.

But two future kingdoms have already fused. Scotland and England have joined in an act of union. Moist rocks have long been home to sheets of green algae, and the first primitive complex plants have begun to grow upright to gather more sunlight and shade the land, and huge water scorpions have begun a tentative invasion of the rivers and estuaries; but as the Devonian begins, there are no terrestrial vertebrates anywhere in this world. Animals with backbones have yet to forsake the sea and try their luck on dry land. Insects have yet to take flight. The little creatures which inadvertently, and capriciously, inscribed this extraordinary story were marine fossils – trilobites, graptolites, fish, shellfish, sharks, rays and so on – that died and left their ghostly shapes preserved to harden in the ancient mud: mute calling cards from an unknown era, that would wait another four hundred million years to be read.

This is the real marvel of the United Kingdom: that this story of the five-hundred-million-year accident of geophysics, geology and evolution should begin to reveal itself in one tiny part of the globe, one of the last to be permanently colonised by humans of any kind, and it should happen during an episode of history called the Scottish Enlightenment. During the seventeenth century, a King of Scotland became at the same time King of England, but the formal union of Scotland and England into the United Kingdom of Great Britain (it would be another ninety-three years before Ireland was included) took place in 1707. It happened at the close of a long period of religious division, bitterness and bloodshed; persecution, witch-burning, civil war and constitutional upheaval all over Europe. Dissenting Christian orthodoxy of the period maintained that the Holy Scriptures were an accurate record of the history of the planet, and one influential and scholarly Archbishop of Armagh, James Ussher, had already computed the chronology of the Bible and compiled a precise timeline. Creation had begun on Sunday, 23 October 4004 BC, when God had said 'Let there be light.' Adam and Eve had been driven from the Garden of Eden on Monday, 10 November of that year, and Noah's Ark had come to rest on Mount Ararat in 2348 BC, as the great flood, product of rain for forty days and forty nights, finally began to ebb away.

But at the same time other devout people had begun to appreciate that even if the Bible was the word of God, the Creator had also written another text that could be read, translated, argued, annotated and admired: the Earth itself was God's handiwork, and therefore bore undiscovered evidence of His craftsmanship. Natural theology was not a new thing – St Augustine of Hippo had proposed something comparable 1,200 years before – but it offered fresh delight for enquiring minds, and all over western Europe minds had begun to enquire. By

1707, Isaac Newton had published his great work *Opticks*, and proposed a theory of light; Halley had predicted a return of the comet of 1682; Newcomen had built his first steam engine; and John Ray had coined the word species and commenced the classification of plants and animals, and in 1691 published a work called *The Wisdom of God Manifested in the Works of His Creation*. The next step, of course, was to begin to wonder how marine fossils could have found their way to rock strata high above the sea. A Danish anatomist called Niels Stensen (usually now referred to as Steno) had already suggested that the fossiliferous strata he found in Tuscany could be explained as deposition after the universal Deluge. But Ray, and other natural philosophers could not see how an event as brief as Noah's Flood could possibly explain all that they, and Steno, had observed.

Science doesn't begin with answers, but with questions: in effect, systematic geology was born, and the first thing the first geologists began to observe was that what they saw in rocks must have required a rather longer interval than the constrained timetable proposed by Archbishop Ussher. It was with such questions in mind that the rationalists of the eighteenth century examined the evidence of the ground beneath their feet, and forged a theory of geology. We fast-forward a bit here: geology turned out to be as tricky as any science, and required investigation of a wide variety of evidence. But gradually the founders of geology made a number of logical conclusions. One was that the deepest sedimentary rocks must be older than the ones nearest the surface; another was that sedimentary strata that were not horizontal must once have been laid flat and then disturbed; and a third important discovery was that the kinds of fossil to be found in certain strata would turn up, again and again, in similar sandstones and limestones in other places, but would be quite different from the preserved creatures above and below those strata. These discoveries were not a challenge to the story outlined in the Bible: they merely suggested that words such as 'day' in the Book of Genesis need not be interpreted as intervals of twenty-four hours. Instead, such words were used to describe intervals of time that could be of any length.* The same observers began to do something else that was to become important: they looked at how rivers deposited silt to estuaries and beaches, and began to calculate the rate at which granites might erode and sandstones and shales might form.

One of the most important of these pioneers was an Edinburgh farmer and physician called James Hutton, and he formulated the argument that processes that happened now had always happened, that the present was a key to the past, and that the paths of rivers, the shapes of mountains, the succession of rocks and the preservation of fossils could all be explained by the slow, uniform action of time, wind, water and processes driven by heat within the Earth. He studied rock formations around Edinburgh, and in the Cairngorms, in Galloway and on the Isle of Arran, and observed some of them to be 'broken, twisted and confounded' as if by subterranean heat and sudden expansion. He looked at the way underlying granite had pene-

---

*Churchmen were among the first to join in the fun. In 1882, in his book *Moses and Geology: The Harmony of the Bible with Science*, Samuel Kinns, PhD, FRAS points out cases in which, in the Book of Genesis, the Hebrew word for 'day' could also mean six days. In Deuteronomy, when Moses says 'Hear, O Israel, thou art to pass over Jordan this day,' he could not possibly have meant within twenty-four hours, because he knew Israel would not enter the promised land until after his death. In the Psalms, a reference to the 'day of temptation in the wilderness' clearly referred to a period of forty years. 'One word more,' says Kinns: 'if by a day we were to understand from sunset to sunset, as observed by the Jews, then how variable such a day would be in different parts of the world, for there is only one day and one night at each of the poles, where the sun shines for six months without setting, and after it has set six months pass away before it again rises.' He then clinches the matter: 'This, I think, quite sets the matter at rest both Biblically and scientifically; and doubtless the whole may be summed up in the expressive words, "One day is with the Lord as a thousand years, and a thousand years as one day."'

trated marble, and must therefore be a younger mineral. He worked out that there was nothing new under the Sun, and

> That before the present land was made, there had subsisted a world composed of sea and land, in which were tides and currents, with such operations at the bottom of the sea as now take place. And, Lastly, That while the present land was forming at the bottom of the ocean, the former land maintained plants and animals; at least the sea was then inhabited by animals, in a similar manner as it is at present.

He saw only, he wrote, 'a succession of worlds', and argued that it would be in vain to look for a beginning of things: 'The result therefore of this philosophical enquiry is, that we find no vestige of a beginning – no prospect of an end.'

It wasn't, at the time, a clinching argument. Hutton was later to be recognised as 'the father of geology', but earth science is a discipline with many heroes, and many arguments, and the story of how Scotland crossed an ancient, vanished ocean and attached itself to what would become England took more than two hundred years of research, argument and more research, worldwide, by tens of thousands of scientists. They began with geological hammers and ended up with space-based instruments; they studied canal beds and road cuttings, examined quarries and walked across deserts. They examined, measured, published and bickered about strata and time scale on six continents and hundreds of islands. They drilled the seabed and mapped the distribution of seismic waves from earthquakes and explosions. They measured elemental isotopes in mineral formations and the magnetic orientation of the iron within basalt rocks. They prospected for gold, silver, uranium, platinum, coal, oil and methane gas, and as they did so they recorded the igneous rocks, the metamorphosed minerals, the sediments and the fossils found in them. They unearthed iguanodons and prised plesiosaurs from the cliffs, and created taxonomies of trilobites and ammonites and graptolites and oolites. They founded societies, published journals and invented a new language, rich in terms such as plutonic and evaporite; horst and anticline; gabbro and orogenesis; strike-slip fault and ignimbrite; albedo and mass extinction; dinosaur and stromatolite; magnetic anomaly and ophiolite; quaternary and zircon; and of course Ordovician, Devonian and so on. Some of them noticed more than a hundred years ago that the outlines of West Africa and eastern South America, and the geology the two continents shared at certain margins, seemed to suggest that they might once have nested together, spoon fashion; but a convincing explanation of how such a thing might have happened was established only in the 1960s.

Earth science involves researchers everywhere on Earth, and lessons from the other planets, moons and asteroids as well, but the story of modern geology really does begin in the Scottish borders; and geology's bedrock – the succession of Cambrian, Ordovician, Silurian, Devonian and Permian eras that span the Palaeozoic – was established by the curious, methodical and very argumentative Scots and English and Welsh and Irish engineers, natural philosophers and amateurs who took up the questions raised, and the propositions formulated, by James Hutton. They studied the chemistry and structure of the rocks themselves, they examined and classified the fossils found in the sediments, they examined formations and identified discontinuities, mysterious places where strata had been folded, even turned upside down, and then eroded before new deposition began; they outlined theories of formation and engaged in ferocious disputes about the triggers for these formations; they learned to recognise the patterns of windblown dunes and tidal ripples that would leave an imprint, millions of years later, in exposed sandstones, and the precise layering of shales that would one day metamorphose into slates; they worked out which lumps of magma had tumbled blazing into the sea and cooled instantly, to become tell-tale pillow lavas, and which had cooled slowly to form characteristic hexagonal pillars; and they learned to identify which rock formations might yield ores, which might conceal fossils tiny or vast. In the course of all this, they did more than discover the secrets of the Earth: they had discovered time as well, a span of time so enormous that even they could at first hardly believe their own speculations.

A society that had more or less without question accepted a biblical story that went directly in a procession of births, marriages and deaths from Adam and Eve to Adam Smith and the Scottish Enlightenment, began thinking in tens of thousands of years, and then hundreds of thousands, and then reluctantly in millions, and at every point in their excavations of the rocks they found evidence of a succession of bygone worlds, of ancient storms and floods and drought and primeval sea that had overtaken a preceding world, and would itself be extinguished. It was this slow confirmation of the sheer depth of time that helped Charles Darwin – until 1859 much more widely admired for his geological research than anything else – to formulate the arguments that turned into his book *On the Origin of Species*. So the intellectual adventure that began in Edinburgh in the century of the Act of Union led to not one but two great, over-arching scientific ideas, and ultimately led to the unravelling of the great story of the act of union with which we began this chapter.

The same approach, the same tools and the same community had begun to explore not just what the United Kingdom gained from the ancient seas, but what it had lost to the latest sea level rise. Within the last two million years, the shape of the united kingdom started to change yet again. London and the Thames Valley and the Essex Marshes and East Anglia seemingly emerged from the sea as temperatures fell and massive sheets of ice more than a thousand metres in height began to advance from the north, planing hills, scouring valleys, grinding rocks to the consistency of flour that would one day become the clay that could be baked into brick. As the north of Britain disappeared under the last great sheet of ice 300,000 years ago, new terrain emerged to the south and the east, and any human bands of hunter-gatherers that foraged in western Europe had access to open savannah and steppe that stretched across what is now the North Sea, and much of the English Channel. Not for the first time, England was an integral part of Europe, and some hardy hunter from the Danish peninsula could have walked across a wide, open plain punctuated with lakes and rivers to visit what would later become Danelaw. So much has been known for decades: simple calculations of sea-level rise and fall, studies of coastal sedimentation, and the examination of bedrock on both coasts, have confirmed that the English Channel opened only relatively recently, separating the Gauls of Britannia from the Gauls of Gallia.

It is one thing to conjecture on an intellectual basis, and quite another to 'see' the drowned lands under the waves that wash the coasts of Britain. But, thanks to high-resolution sonar technology and the evidence from seismic soundings, it is possible to reconstruct a vanished landscape. Imagine, at the height of the last Ice Age, a mere 20,000 years ago, a huge valley, a floodplain, extending from the southern North Sea all the way down the English Channel, past the Lizard of Cornwall, past the westernmost capes of Brittany, before it halts at the commencement of the deep Atlantic. This is now the European continental shelf: once it was almost all part of Europe proper. Down it, racing westwards, surged a vast, wide river, to be joined at its westernmost reach by a series of other little rivers from the hills of Brittany and Devon. The Channel river is enormous: naturally, because its tributaries are the Rhine, the Seine, the Scheldt, the Thames, the Meuse, the Rother and the Solent: seven great rivers feed their burden into one super-waterway. It represents the ultimate drain on western Europe's resources; it carries melt-water, silt and dead foliage all the way from the Alps and the Cotswolds to the sea along one vast, meandering, swollen, island-strewn channel, and at certain points and seasons each year it must have been as much a challenge to cross as the Channel is now.

The North Sea, too, off Spurn Head and the Wash, must have been open landscape, cut by meandering streams and dotted with stagnant shallow lakes. From time to time (palaegeologists have identified at least two such moments) these landscapes were scoured by colossal floods: huge bodies of water would build up in temporary dams that would ultimately burst, and sweep across the low-lying ground with devastating consequences for any living creature in their path. These tracts of now-submarine landscape were once inhabited: North Sea fishermen and gravel dredgers have over the years brought up more than a hundred Neanderthal flint axes, along with Stone-Age tools fashioned from flint, bone and antler, and the remains of mammoths, rhinos and other big game that roamed the Palaeolithic savannah. Along the coast of Norfolk there was, 700,000 years ago, a huge forest, and in the bed of a river that once flowed through it have been found thirty-two flints that show every sign of having been chipped, hammered or flaked by visiting humans. The pollens, fossilised plant tissue and insects found in the ancient forest beds suggest that for a time the region enjoyed a Mediterranean climate, with a succession of advancing and receding seas lapping the ancient coastline. But Britain was then a peninsula rather than an island (it would be cut off by rising seas on occasion, but only relatively briefly, during the next 700,000 years). The summers were hot and the

winters mild and moist, which would explain why hippopota-
mus were comfortable enough there, along with bison, the
extinct mammoth, rhino and giant Irish elk. Big game attracts
big predators, so lions and spotted hyenas also stalked the
grassy plains or lay concealed in the forests of oak and
hornbeam.

The humans who walked into this vanished landscape and
fished flints from the gravelly riverbed to make tools would not
have been *Homo sapiens*. Anatomically modern humans probably emerged in Africa only 250,000 years ago. England, if it was
'home' to anybody, was part of the range of a long-extinct
species called *Homo heidelbergensis*, forerunner of the
Neanderthals and *Homo sapiens*, and it was probably a party of
H. heidelbergensis that 500,000 years ago crossed the causeway
that connected England and France, set up a little camp at
Boxgrove in West Sussex, and left calling cards of stone: a pattern
of chipped flints, some rhinoceros bones that had been cut from
a carcass and broken open for the marrow, a human shinbone
and one human tooth. At the same site, archaeologists found
ancient evidence of lions, leopards, hyenas, bears, giant deer, red
deer, bison, elephants, badgers, hedgehogs, spoonbills, mallards,
greylag geese, eagle owls, snakes, frogs and toads. Finds such as
these are serendipitous, and tantalising. It is in the nature of
biological tissue to decay into dust and ash and air and water: to
be preserved in fossil form, teeth, bones, feathers, skin and other
tissue must be buried suddenly, ideally in mud that will not
permit ordinary decay, with more and more material piled on
top that turns wet sediment ultimately into rock. In the long
run, all the biological tissue will vanish, to be replaced entirely
by mineral material, a molecule at a time, but this process will
sometimes preserve not just the shape, but even the layering of
flesh and bone and sinew originally buried in a mudslide or a
huge fall of volcanic ash.

But such things happen haphazardly, and hardly at all, and even when they do happen, the fossil remains are at the mercy of time's caprice. Rocks once formed may be crumpled by tectonic uplift, ground by glaciers, dissolved by rain, broken by frost and abraded by sandstorm. Once exposed and eroded, any fossils exposed will be eroded with them. So serendipity has to be at work a second time: observers must be around for just those few months or years in which the fossil becomes apparent in the rock, and one of those observers must be savvy enough to recognise the fossil for what it is, and then responsible enough to alert the archaeologists. This process explains why fossils are rare enough to become collectors' trophies, and why human fossils are even more rare. Human fossils in Britain are so rare that scientists in 2001 launched a project called the Ancient Human Occupation of Britain, because for one 100,000-year stretch during the last Ice Age there didn't seem to be any humans in Britain at all. Clearly, an earlier species of Briton migrated from Europe and stayed for the hunting season more than half a million years ago. As the climate cooled 450,000 years ago, the glaciers began to pile up and the seas receded, it became even easier to get into the united kingdom: easier, but in terms of good hunting, perhaps less rewarding.

From 700,000 years ago to about 150,000 years ago, there are little pockets of evidence of human settlement, and even a few human bones. But for the next 100,000 years there is no evidence of any human. This may be linked with the last Ice Age: glaciers covered the high hills; what had been woodland became tundra; hippo and elephant retreated and their places were taken by mammoth and woolly rhinoceros. At intervals, the cold relented, and pine and spruce grew where there had once been oak and hornbeam, and then the ice moved south again. The final cooling ceased abruptly ten thousand years ago; the ice disappeared from Britain and northern Europe; the ocean began to reclaim the low-lying plains between Norfolk and Jutland, the Rhine and Meuse began to spill into the North Sea and the Channel River began to widen into what would become the Channel.

With this most recent dramatic rise in temperature, the united kingdom has become an island state, inhabited by modern humans with wooden-shafted spears, stone axes, flint blades and needles of bone and clothes fashioned from fur and hide, and a taste for decorative jewellery. Five hundred million years have passed in the construction of what will become a sovereign state, and absolutely everything that we think of as history – the bronze and iron ages, Druid rituals, Stonehenge, agriculture, civil engineering, wheeled traffic, the domestication of animals, the confection of wine, bread and cheese, the creation of villages, towns and cities, the invention of writing, the keeping of calendars, the forging of epic verse and ritual theatre, the formalisation of religion, the coronation of kings, the philosophy of government, Shakespeare, the European discovery of America, the industrial revolution, political revolution, the birth of modern science, and the slow, painstaking exploration of the fabric of the planet – will all happen in the next ten thousand years. And 9,700 of those years will pass before the formal declaration of the first Act of Union that will launch the United Kingdom.

SIX

The Continent

An Attempt to Join Europe

The word 'Europe' is not in my *Shorter Oxford English Dictionary*. It occurs in the full twenty-four-volume *Oxford English Dictionary* (second edition) only as a term used 'elliptically and allusively' since 1957 for a political marriage of convenience known as 'Europe' and from which, for many decades, most European nations were excluded. The word European, according to the *Oxford Dictionary of Etymology*, derives from a Greek word of unknown origin, and its meaning was applied at first to central Greece, and then to the whole Greek mainland, and finally to the landmass beyond. The first use of the word European, to describe the people who inhabit the continent, happens when a scribe refers to victory by the 'Europenses' in the battle of Poitiers in AD 754, against Islamic forces. So the term European becomes one of distinction: to distinguish one set of humans from others who are therefore not Europeans.

As a precise geographical address, Europe is easy enough to point to on a map, but rather harder to define. Where does it begin? Where does it end? Is Europe actually a continent? Once again, the vagueness begins with the word itself: continent's first meaning is a containing agent or space, according to the *OED*, which of course explains interesting variants such as incontinent. The fourth meaning of the noun covers mainland, as opposed to islands: but this mainland doesn't have to be very big. England is a continent, compared to the Isle of Wight, according to a use recorded in 1628. The notion of Europe as the Continent (as distinct from the British Isles) is however an old one: 'the continent, where everie kingdome and state doth joyne one to another without anie partition of sea', observes a writer in 1590. So, in one definition at least, and for some of the citizens of just one nation, Europe begins at Calais. That still raises the question of where Europe ends. Geographers might define a region loosely bounded by the Arctic Circle, the Urals, the Caucasus, the Black Sea, the Mediterranean and the Orkney islands, tied by a long history of human migration, conflict, trade and treaty. Russians, according to mood and historical experience, would argue about whether they were in or of 'Europe'. Politicians would include Iceland and perhaps Greenland but argue about Turkey. Historians would add the legacy of religion, empire and culture as part of the embrace that makes Europe an entity. Greek civilisation, for instance, may have its roots in Asia Minor, and certainly spread far beyond Greece: Naples in Italy is a corruption of Neopolis, or New Town. Marseille in France derives from its Greek name Massalia. Most of mainland Europe was once under Roman rule. So there is something in history that unites Europe: Greek intellectual tradition and Roman imperium.

That still leaves a number of loose ends. Scandinavia, the Baltic countries and Scotland never submitted to Roman sway. On the other hand, the Romans for a while held territory all the way to the Euphrates, and Egypt, Libya and Carthage as well.

Is Europe the fabrication of a loose, unruly Christian congregation? Is there a core Europe, a historic centre united in Christian faith, which defines our notion of the European continent?

If so, then the story begins far from what is now notionally Christian Europe. It begins with Saul of Tarsus, a Jew from what is now southern Turkey who rides east to persecute a small group of Jewish dissidents who are now known as Christians; who has a change of mind, heart and name on the way to Damascus in Syria; and who, as the Apostle Paul, does most of his proselytising in Turkey and Greece before being shipwrecked on Malta and imprisoned in Rome. It continues with the story of the young soldier Constantine, born in what is now Serbia, who at a critical moment in imperial history is quartered in Eboracum, the modern city of York in Yorkshire; who heads eastward to Italy to begin a bloody civil war; who claims the title of emperor of Rome; who then shifts the capital of the greatest empire the world has ever seen to new premises on the Bosphorus, which he names after himself, using words from the Greek; and who declares Christianity the official religion of his empire while himself remaining – until his deathbed conversion in a city in what is now Turkey – officially pagan.

The greatest players in this story often come from the periphery: the African-born St Augustine, perhaps the son of a Berber, laid some of the foundations for the tradition of European thought from an unstable intellectual power base at Hippo Regius in North Africa (just before the Vandals sacked the place, just as they sacked Rome). Europe begins to look like an adventure playground for people who were not necessarily born Europeans: Huns and Goths and Burgundians; Magyars and Slavs and Scythians; Crimean Tartars and Ottoman Turks and Saracens and fierce clansmen from the Scandinavian fjords who over the centuries became the Varanger guard of the emperor of Constantinople; the merchant-burghers of York and Dublin; the nobles of Normandy and the kings of Sicily.

Europe sometimes seems like the ultimate airport novel: the plot is confused, the action is relentless, the setting is inexplicit, and – quite often – the characters have so little substance that to call them cardboard would be an insult to cardboard. Who were the Huns, and from where did they begin their expansion westward? Who were Gepids, the Alani, the Visigoths and Ostrogoths, the Lombards and the Burgundians? Where were the boundaries of Neustria, and Austrasie? What happened to Visigothic Septimania? Why was there a Regent of Brunswick but an Elector of Hanover? Who were Clovis I, and Childebert II and Clothar III? Who was Pippin III, also known as the Short, who became King of the Franks and thereafter known as Pepin the Short in AD 751? Some of these questions can be answered in a few sentences, with reliable references, and some cannot. Who was Rollo the Viking, who was handed control of what is now Normandy by Charles the Simple? Was he a Norwegian? Or a Dane? How did he become Robert, and was he a duke, or a count?

Everything in Europe seems to have come from somewhere else in Europe, and before that, from somewhere else beyond the European subcontinent. This list of alien intruders includes the first bearer of the name itself, the European alphabet, the number system, the calendar, almost all of Europe's mammal livestock, and that pan-European signature lunch, the ham sandwich. Europa, according to the earliest myths, was the Phoenician goddess, or perhaps princess, abducted by Zeus, who adopted the guise of a white bull and swam away with his prize to Crete. Homer even has Zeus admitting the rape, in Book 14 of *The Iliad*. Europa also occurs in Greek mythology as one of the daughters from the marriage of Oceanus and Tethys. She is invoked by the Greeks, but Greek legend also has it that the peninsula famed for Mycenae and Argos was settled by, and takes its name from, Pelops of Phrygia, a kingdom which is now part of Anatolia in Turkey; Cecrops, founder of Athens, is supposed to have migrated from Egypt on a civilising mission.

The European alphabet most certainly comes from beyond the shores of Europe: it takes its name from alpha and beta, the first two letters of our ABC, but the Greeks based their lettering system on a Semitic script: Phoenician, or proto-Canaanite, which is also the origin of the Arabic and Aramaic, so the alpha and beta of New Testament Greek is first cousin to the aleph and bet of Old Testament Hebrew.

The numbers by which the world now counts were gratefully adopted by Europeans from the Arabs who invaded the Mediterranean region after the seventh century AD (I used to lie awake at night wondering how the Romans did multiplication and long division and fractions with a system that involved only the letters I, V, X, C, M and so on), and the Arabian mathematicians had in their turn improved a number system that may have had its origins in the Indus valley, or in India.

The division of the day into twenty-four hours and the circle into 360 degrees and the division of the stars in the heavens into twelve constellations of the zodiac all begin somewhere in the land of the two rivers, the Tigris and the Euphrates. So, for that matter, does most of what we count as European civilisation. If we are what we eat, then our European diet begins in the Fertile Crescent. This is an old name for a wonderful semicircle of ancient cities, settlements and estates that extends around the north of the Arabian peninsula, running from the mouths of the Tigris and the Euphrates at the northern end of the Persian Gulf, around the Taurus mountains of modern Turkey and then down through northern Syria to the shores of the Levant and ancient Palestine on the Mediterranean. Some maps extend the crescent all the way to the Nile delta in Egypt. It is a landscape of place names that sound like fragments of an old epic: Babylon and Nineveh and Sidon and Tyre; Carcemish and Sumer and Akkadia and Nimrud. Jerusalem is there, along with Jericho, Damascus, and Byblos, too, the city that gave us a name for books and bookishness, and for what Puritans used to call the Good Book.

It is also the place where agriculture was born. Eleven thousand years ago, at the end of the Ice Age, some wandering party of Neolithic hunter-gatherers began to harvest and store an edible flower called the fig: fossil specimens have been unearthed not far from what would become Jericho. *Ficus carica* is a useful discovery: its leaves provide shade, its fruits are plentiful and they can be dried, to become even sweeter. Figs dried or fresh would have provided portable calorific energy, along with a range of valuable minerals, vitamins and anti-oxidants; a packed lunch, so to speak, for the journey towards settlement. The fig was only the beginning: the same sickle-shaped garden of water and rich soils was home to other wild plants ripe for the taking and sowing. Two of them – emmer and einkorn – became the ancestors of modern cultivated wheat. Barley, too, became a staple, seasonal crop for the first people who decided that it might be better to stay in one place, and store food through the winter, first as a grain with which to make gruel or cakes, and then much later as the basis of beer.

Another useful plant that first grew wild in the Fertile Crescent was the chickpea, a third the lentil, a fourth the pea, a fifth the faba or broad bean. The flax plant was also first harvested in the crescent: its seed serves as food, its fibres became the basis of linen. It was within this zone of possibilities that humans first domesticated cattle, from the aurochs or wild bovid; first rounded up the wild sheep and made it a source of meat, milk, cheese and fleece; first began herding the wild goat. Somewhere in this landscape, too, Neolithic settlers began to harvest the calorie-rich olive. Food is the first source of human power, a fuel, a source of energy that drives survival and replication, and if it can be gathered and stored it becomes the foundation first of civic settlement, and then civilisation. Homer's heroes counted their wealth in jars of oil. Olive oil burns, it flavours, it serves as an unguent, but it also nourishes, and soon enough it acquired ritual value. Bronze Age kings and high priests were anointed with olive oil; heroes were massaged with it. Its branch became the symbol of peace, its oil the fuel for a sacred flame.

Grapes were perhaps first exploited somewhere beyond this region, but the first fossil seeds date from the beginning of the Bronze Age, near ancient Jericho, and Egyptian farmers knew how to make wine more than four thousand years ago. Dried grapes make portable, easily stored food; fermented grape juice, too, is an important source of calories in the winter months; mixed with water, it made for a safe drink; it too now has religious connections that must be far older than the Christian connection of communion with bread and wine. Homer's heroes make libations with wine, Greeks celebrated the god Dionysus with Bacchic rituals that involved heroic drinking, and the wonderful, infectious quality of enthusiasm derives from a Greek term meaning 'divinely drunk'. Wheaten bread and barley meal and beer, beans and peas and lentils, olives and olive oil, raisins and wine, beef and mutton; butter, cheese, milk and yoghurt, all have their origins in the Fertile Crescent.

Of all the staples of the modern European diet, only the pig is truly European in origin: that is, Europe's pigs count their descent from the European wild boar. But even this is an illusory innovation. The pig, too, was first domesticated from a wild progenitor somewhere in the forests and fertile soils south of the Taurus mountains, and the first pigs to be introduced to Europe were from the Middle East. In 2007, researchers led by a team from the University of Durham sampled the DNA from hundreds of preserved pig bones and teeth and calculated that the first piggeries in the Paris basin five or six thousand years ago were stocked by swine bred from stock first imported from the Middle East. Nobody knows why, but pigs with this genetic fingerprint have long since died out in Europe: they were

completely replaced, within a millennium, by the native ancestors of the Large White, Tamworth, Gloucester Old Spot, Wessex
Saddleback and so on; all bred from the European wild boar,
rather than the Gadarene swine of the Bible lands. It turned out
to be a useful discovery: it confirmed that the pig had not been
domesticated independently in both places, and then if not the
pig, then probably not the cow or the sheep, the ancestors of
which must also have ranged the European heartland.

Agricultural practices spread across the Anatolian plateau,
and around the Black Sea, slowly along the Danube basin, and
then around or over the Alps to the Rhine and the Rhône, to
France and Germany, and surely, too, along the Mediterranean
coastline, all the way to Iberia and then up to Paris and the
Channel coast. Perhaps some of the pioneers who brought the
spade and the hoe were people from the Middle East, or their
descendants. Europe's first ground-breakers certainly seem to
have been migrants: in 2009, European scientists analysed the
mitochondrial DNA – which is inherited only from the mother,
and is therefore a useful way of tracing family connections –
from the bones of early European hunter-gatherers, and
compared the results with samples of the same genetic material
gathered from the remains of the first farmers of north and
central Europe, 7,500 years ago. They did not match, according
to a report in the journal *Science*. The first farmers to turn their
hand to the plough, at least metaphorically, in Poland and
Prussia and the Rhineland of the Palatinate were not descendants of the first hunters and fisherfolk who had set up camp
after the retreat of the glaciers. There must have been some cases
in which hungry hunters and gatherers saw what their immediate neighbours to the east had provided for themselves, and
learned from them. But the first innovators were incomers: they
were migrants. The debate is not over, and no one yet claims to
know where the first German pastoralists and planters arrived

from, but they certainly came from the east, because they brought with them the flocks and herds, the digging tools and the pruning knives, the grains and the pulses and the vines and fruits first domesticated in the east, and along with them trotted the pig of the Fertile Crescent.

So even though the pig is anathematised in the lands now dominated by Islam, and pronounced unclean in three-thousand-year-old Jewish scriptures, the Middle East was the primary provider of all those things now considered part of the great West European tradition: bacon and ham and salami and sausage; bread and cheese and baked beans; beer and wine; oil and vinegar. Chuck in a few recent imports: spices from the East Indies, potatoes, tomatoes, chillies and corn on the cob from the Americas, and the Europeans are left with very little on the table that is essentially West European. Kippers, anyone? The horse, too, came cantering in from the steppes: there is evidence that mares were being milked, and horses bridled and then driven or ridden, in Kazakhstan and Ukraine more than five thousand years ago. Wild horses existed in Europe at the end of the last Ice Age: their outlines are beautifully preserved in Pyrenean cave paintings, and they were hunted like deer and wild cattle all through the Old Stone Age. But the idea of the horse as an extension of human range, as a tool of superior mobility, as a form of horsepower, began somewhere else. Chivalry, that medieval European ideal, rode in with the cavalry of the east.

One has to marvel at the mix of foresight, ingenuity, cooperation, favourable circumstances, willingness to experiment, and above all patience that led to the separate inventions of agriculture, both arable and pastoral. Docility is a feature of most modern dairy herds. Such wild cattle as survive in Africa are not docile: the Cape buffalo is reputedly a very dangerous customer, and the now-vanished aurochs, ancestor of the Hereford and Aberdeen Angus, Sussex Red and Charolais, Friesian and Jersey, must have been able to throw its weight around in a landscape stalked by lions and wolves. Sheep now graze peacefully and move relatively slowly, the consequence of seven thousand years of selective breeding for meat, wool and milk: wild sheep and goats however can move like big cats. Hunter-gatherers would have already known and appreciated the nutritional value of certain wild grains: but it would take a certain faith in the future to encourage a group of seasonal nomads accustomed to hand-to-mouth existence to settle instead in one place, and wait for uneaten seeds, strewn at the end of winter, to sprout, grow, flower and ripen. To make settlement secure, the first agricultur-alists had to invent irrigation, pest and weed control and muck-spreading; they had to design ever more efficient stone tools and digging techniques; they had to build secure storehouses proof against insect and rodent infestation; they had to start experi-menting with ways of grinding grain into meal, and then into flour; they had to fashion receptacles, cooking techniques and ways of preserving and storing the surplus and – since this surplus represented survival, or wealth, or both – guarding or protecting it. It isn't then hard to imagine some individuals, or families, or even tribes, specialising in particular forms of production, which would in turn entail distribution and exchange.

These ancient decisions live on. Crafts such as smith and carter, cooper and carpenter, miller and potter, wright and ward, hunter and fisher, chapman and shepherd, cook and baker, have all become common English surnames, and have their parallels in many languages. It isn't hard to see how civilisation – in the most practical sense of the term – might have spread. A technol-ogy, an artefact, a skill or a food source would have been observed and adopted, by one community after another, in a steady progress from east to west; in return, the westernmost communities would have adapted such things to their precise circumstances, and fed back the most useful improvements down the chain to their source.

Commerce would have been an inevitable part of this traffic: pork has even more value if it can be smoked or salted and kept over the winter as bacon, or ham, or sausage: the first is a specialist skill that must be studied; the second requires salt by the bucketful. So a little of the substrate of Europeanness – bread and beer, knives and casks, pickled vegetables and dried fruit, bacon and beans – spread everywhere on the continent: it reached villages built on stilts in the Swiss lakes, and tribes that lived by fishing on the Danube, and the people of the forests of the north. Milk proteins have been found sticking to bits of pottery from what is now Hungary and Romania, that date from 7,500 years ago; smears of milk fats turn up in ancient British archaeological remnants dating from around six thousand years ago. Where bread and cheese and bacon went, sooner or later the knife and fork followed. Across the Turkish mainland and over the Bosphorus, and up the Romanian coast of the Black Sea to the Danube came the first artefacts of bronze, and later of iron. Boats of a sort must have existed wherever there was water, but the people who made bronze needed tin from Cornwall as well as copper from Cyprus, so sturdy sailing craft accelerated the spread of civilisation around the Mediterranean coast, and the people most famous for making capital from a sustained programme of marine investment were the Phoenicians.

So a process that began around 12,000 years ago in the Fertile Crescent, little by little, sometimes peacefully, sometimes violently, provided the foundations for the communities to the west that became tribes, races and nations, with their own traditions, races and cultures. Natural fermentation is a process that begins when a tiny speck of wild yeast lands in a still mush of biological substance: just introduce a potent alien idea, or entity, or force, and new, nourishing and intoxicating things happen, and go on happening.

It is out of such transmissions that this many-headed, and therefore difficult to define, entity called Europe came to be. What we think of as Europe is influenced by the earliest memories of things we saw and learned. As an altar boy in a small, immensely quiet New Zealand suburb, I learned fragments of Latin quite casually, as if it were a normal thing to do. *Introibo ad altare Dei*, the priest would intone, and I would answer, *Ad Deum qui laetificat juventutem meam*. 'I will go to the altar of God ... to God the joy of my youth,' the invocation and response read in the English pages of the missal, so at some imprecise age before my tenth birthday, I knew that my language contained words – deity, juvenile, me, altar and entry – that had histories that went back to Julius Caesar, and Constantine, and Augustine and Gregory the Great; and to some confused period that followed – historians called them the Dark Ages, and the Middle Ages – where you could notionally be safe from arrest, torture and murder in a sanctuary, that separate place in a church beyond the altar rail. Somewhere in this arbitrary outpost of Europe, this new world settled by people who really had wanted to 'start again', by people from England and Scotland and Ireland, from Holland and the Dalmatian coast of Yugoslavia, I understood that oil, and wine, and bread, and salt had powerful ritual significance that dated back far beyond the Romans, to some Bronze Age pastoral peoples of the Levant. Much, much later, it was with a thrill that I learned from Victorian travellers such as C.M. Doughty, author of that lonely marvel *Arabia Deserta*, that in the Hejaz and along the pilgrim roads, fellow travellers were obliged to defend and support each other because they had shared bread and salt: that was all it took to make a true companion (the Latin word itself contains the clue); and that Russian cosmonauts aboard the space station *Mir* welcomed the first American astronaut visitors with the same ritual of bread and salt.

One advantage of a Catholic education was that it provided a second, coincident value to words that survived in the secular world: doctrine, dogma, propaganda were attached dismissively and sometimes indiscriminately to prefixes such as Marxist, communist and socialist. But, of course, the difference between doctrine and dogma is that your membership of a faith permits you to question the doctrine but requires you to submit to the dogma; and if you have a faith, what else should you do but try to propagate it? So, in a polarised world, and as an unquestioning member of a huge and influential religious sect committed for at least four decades to a very conservative form of democratic capitalism as the only path to salvation, I understood that Communism too was a faith; I began to learn to see things from another point of view, simply because I began to learn about the real meaning of the words around me.

Catholicism was a story also told in pictures: by Giotto and Michelangelo and Raphael and Caravaggio and Titian and of course by Hollywood – the Hollywood of preposterous entertainments such as *The Robe* and *The Song of Bernadette* – but these works of art existed beyond the boundaries of anything I understood as reality. Very great art, like a lot of very bad art, has no fixed abode. It is neither here nor there. It just is.

The earliest pictures I can remember that specifically evoked a Europe that I might one day see and feel and smell and touch were very different from these. One of them was Jean-François Millet's *The Angelus*. It shows a man and a woman, heads bowed as if in prayer, standing in a potato field, conforming to a ritual announced by the tolling of the Angelus bell. The landscape is flat, and there is the spire of a church in the distance. Hay or corn is stooked in the next field; in the foreground is a pitchfork and some exposed tubers. I understood that they were somewhere on the great European plain, a fertile level that, my father's relief maps told me, extended from the North Sea coast all the way to the Urals.

The other was a cover in a display case outside a small bookshop that was reputed to sell 'fast' literature. Alongside Mickey Spillane's *Kiss Me Deadly* (the cover of which showed a man pointing a gun at a girl who was unbuttoning her blouse) was a novel called *Savage Paris*, the dust-jacket illustration of which also appeared to evoke erotic violence: a buxom, full-bosomed young woman holding a bloodied knife and staring scornfully at a thin, unhappy man in uniform. It was a 1955 English translation of Émile Zola's *Le Ventre de Paris*, written in 1873 and set in the food markets, but I wasn't to know that for some years.

These two images defined the France of my youthful imagination: a place where in the countryside you could see people lifting potatoes, and in the city you could see them buying and selling food. But they also helped me at least begin to understand the idea of Europe: the intricate set of connections that stretch across national and regional borders, and even across oceans. They became, although I could hardly have been aware of this at the time, unconscious illustrations of the history of a continent.

At around this time, I confronted history, along with geography, as a secondary-school matriculation subject: it was, of course, British and European history, involving tales of Robespierre and Napoleon, Talleyrand and Metternich and the Congress of Vienna, and the great European narrative seemed to come to a halt with Bismarck and the reunification of Germany and Garibaldi's march on Rome. But these were just names, dates, sequences of events. We have memory, and we have imagination, and for much of our lives these are quite arbitrarily detached from each other. Every now and then something will happen to skewer that detached awareness of an item, a

figure, or an event to a real landscape or address: then the imagination can start to work. Some bits of European history fell into
place long, long after I first read them. One such placing, for
instance, was the discovery of the existence of a Mazzini-
Garibaldi Club in London, near Red Lion Square. It has now
vanished, but at the time it was within easy walking distance of
the Marx Memorial Library, on Clerkenwell Green, from which
Lenin edited his journal *Iskra*, 'The Spark'. So this was another
pleasing reminder of the manner in which movement defines
the continent to which I owe my tradition, my education, and
my cast of mind. People crossed borders, joined communities,
confected ideas, created ferments, transported enthusiasms and
promoted revolutions.

This mobility certainly did not exist in the stillness of Millet's
Angelus or the frozen emotions expressed cartoon-fashion on
the Elek Books cover of *Savage Paris*. But behind the stasis of
each remembered image is a story of perpetual motion and
endless commotion. The people in Millet's picture are alive
because they, and their families before them, could exploit
Solanum tuberosum, a crop that survives in the ground over
winter, and delivers calorific energy in denser packages and
greater quantities than bread, and with less fuel or fuss; a crop
first domesticated somewhere in the Andes in South America.
The potatoes were a consequence, an accidental discovery, which
followed from the obsessive ambitions of a young sailor from
Genoa who accepted Spanish backing to find a new way to the
Orient by sailing westward. Columbus of course found the
Americas instead, and although Britain and France fought for
America north of the Rio Grande, and the Spanish and the
Portuguese divided almost all of the rest of the two newly-
discovered continents between them, the French colonised Haiti
and Martinique and part of Guiana, and that act of colonial
possession provides the link between Millet's *Angelus* and Zola's

Savage Paris. The sad, famished man on the Zola dust jacket has just returned from Devil's Island, to which he had been transported after his arrest during the bloody and cruelly suppressed revolution of 1848. He returns, starving and penniless, to a Paris obsessed by food and wealth, and ironically gets a job as an inspector at Les Halles, where food and wealth are all that matters, and inseparable.

The year 1848, of course, is the year in which Karl Marx published *The Communist Manifesto*, and Marx was a Prussian of Jewish parentage who was born in Trier, who moved to Brussels and Paris and then finally settled in London. The paintings of Millet – in particular his 1857 study *The Gleaners*, which shows three stooping women scavenging the leftovers of the harvest, were considered socially threatening, because they focused on the rural poor; one critic at its first exhibition saw in it 'intimations of the scaffolds of 1793'. Millet had been thinking of the story of the gleaners in the biblical story of Ruth and Boaz, so just two remembered images make a series of links that connect Millet with Zola, Marx with the Bible, plenty with hunger, peasant provender with the European discovery of America, and Catholic devotion with obscene libel, because that was the crime for which Zola's first English translator and publisher, Henry Vizetelly, was prosecuted, fined and imprisoned under English law.

Almost a century on, there was still something dangerous about Zola. His complete works remained on the Vatican's notorious *Index Librorum Prohibitorum* until the abolition of this ridiculous and shameful entity in 1966. The first I knew about these things was when I learned that 'good' books were the ones that had printed in them words such as *Nihil obstat* or *Imprimatur*: the first assures me that a bishop thinks that the book contains nothing contrary to faith or morals, the second that the work is free from error in matters of Catholic doctrine.

The arrogant implication of such terms – the assumption that there were people who knew what was best for the layman, people who could prohibit what ordinary people might read or hear – was not of course confined to the Catholic Church: the US government in the 1950s famously withdrew the passports of some of its most admirable citizens, among them the black singer, actor and political campaigner Paul Robeson; the United Kingdom until 1968 maintained censors within the Lord Chamberlain's Office who controlled what theatregoers might see on the commercial stage. I grew up among people who could remember that the Nazis consigned books to the bonfire, and Ray Bradbury had in 1953 written *Fahrenheit 451*, that wonderful novel about a world purged of books. The invention of the book made possible the most extraordinary companionships across space and time: I can stroll about London with Samuel Pepys or James Boswell, or go on the grand tour with John Evelyn; I can receive letters from Seneca, or the Younger Pliny; or enjoy the imperial court gossip from Gaius Suetonius; go into battle with Patroclus in *The Iliad*; or stay away from the front line with Jaroslav Hašek's Good Soldier Svejk; cruise the tropic oceans with Ishmael and Starbuck aboard the *Pequod*; take the little train from Balbec to Paris with Albertine and the narrator of *The Remembrance of Things Past*. These are intimate pleasures, but ones that I can share with millions, some of them contemporaries of Pliny and Proust, many of them yet to be born. The destruction of the great library of Alexandria at some point in the first millennium deprived posterity forever of an unknown (and unknowable) number of scrolled master-pieces from the Graeco-Roman world. But books printed in their thousands and tens of thousands can never be vulnerable: only one copy needs to escape the bonfire, and it remains a menace, made all the more potent by awareness of attempts to eliminate it.

Books, in a sense, were the making of modern Europe: the printed book reintroduced people of Western Europe to the ancient past with the Renaissance of learning. After that time, all sorts of people could delight in Ovid and Virgil, Homer and Herodotus; they could also print and distribute their own epics of storytelling, the *Heptameron*, the *Decameron*, the *Gesta Romanorum*, the *Golden Legend*; they could focus their arguments and promote their beliefs in the broadsheets, pamphlets and tracts that marked the religious bloodshed of the sixteenth and seventeenth centuries; they could address the practicalities of the political and economic constitution and write books that aimed at sedition, subversion and change. They could challenge the Bible, they could denounce the king, and they could compose comedies and tragedies of romance and scandal and passion and intrigue. Printed books could promote rational discovery: Copernicus, a Polish cleric and civil servant, launched a scientific revolution with a book known appropriately as *De revolutionibus*, and escaped censure by the Church by adroitly presenting his thesis as a mere hypothesis, not necessarily a representation of reality. Galileo Galilei took up the Copernican challenge of *De revolutionibus orbium coelestium*, and presented his evidence in the form of a Dialogue, but this sophistry did not save him: he was convicted and condemned for 'grave suspicion of heresy', and the *Dialogue Concerning the Two Chief World Systems* was placed on the Index for the next two hundred years.

But that is the marvel of Gutenberg's achievement. You can condemn an idea; you can convict and burn a heretic; but you can never quite suppress a book. The ancient author Ptolemy said, authoritatively, that the Sun went round the Earth, and this thesis happened to suit the Church very well, for more than a thousand years, because it put Earth as the centre of creation, the sole focus of heavenly attention. Giordano Bruno and Galileo believed that the Earth went around the Sun and said so, and one of them was burned at the stake and the other forced to recant his heresy; but after publication, neither of the individuals mattered: what mattered were the books. Publication provided somewhere for the next investigator, someone beyond the reach of the Inquisition, to start, to confirm, or to demolish, or to enhance the argument.

Science was driven by the printed word; and so, in different ways, was romance itself. With printed books, people in Europe could formulate the invention of a parallel world of unreason, of improbability: stories of kings, heroes, champions and lonely, heroic knights in battles against extraordinary odds. Stories relayed in pictures, ballads and scrolls were the making of national heroes, but printed books made them international heroes.

Vlad Țepeș, Vlad III, Voivode of Wallachia in what is now southern Romania, is also known, in his own country and far afield, as Vlad the Impaler, a monster of cruelty and caprice. He was all these things to the people of communist Romania, but he was also rather proudly saluted as a local hero: Vlad had been the last Christian prince to resist the Turkish conquerors after the fall of Byzantium in 1453. Sure, he was cruel, I heard Romanian scholars admit airily, but this was medieval Europe in bloody times. The Borgias were cruel. The Inquisition was cruel. The Tudor monarchs were cruel. Vlad Țepeș was a man of his times, and his people respected him for it. You knew where you were with Vlad the Impaler. He was firm but fair. I heard these things at the first-ever international conference of the Transylvanian Society of Dracula, an attempt in 1995 to reconcile the two clashing models of Dracula. Until the fall of the communist regime at the close of 1989, Romanians knew nothing about the movie Dracula, the histrionic vampire played in British horror movies almost always by Christopher Lee, and always foiled in the last reel by Van Helsing, played almost always by Peter Cushing. Westerners thought they knew about the historic Vlad of the house of the Dragon, or Dracul; most of us were not aware until this event that the principality in Vlad's time, and the country now, was home to an uneasy mix of Saxon Germans and Magyars from Hungary, Turkish settlers and people who claimed descent (and certainly spoke something recognisably Latinate) from the legions of Trajan. The Westerners, conversely, were there to discuss a fictional character, a player in an entertainment confected by a late-nineteenth-century theatrical agent from Dublin, who originally set his monstrous story in the Carinthian region of Austria, but then changed it arbitrarily to Transylvania, a country he had never visited.

The novel at first publication enjoyed only modest attention: it was not until the invention of cinema that Dracula became an enduring multi-media hero, and Bram Stoker's book began to look like a classic. So this conference staged a series of academic bunfights in locations important to the real Dracula, and his fictional namesake: Bucharest and Sighisoara, Vlad's birthplace; Bistritz or Bistrita where Stoker's hero Jonathan Harker described his last agreeable meal; and a castle on the Borgo pass, one in which Dracula certainly never lived, because it had been newly purpose-built for Dracula tourism. The organisers did so to celebrate two great European identities: an indestructible monster paradoxically destroyed again and again with a stake through his heart, and a determined warlord who demonstrated his determination by impaling his subjects, prisoners and enemies with a stake through other parts of the anatomy.

The event was a reminder that Europe is not just a collection of countries, nations, languages, cultures and identities, and traditions, linked by mutual histories and celebrated by shared table fare – bread and preserved pig are standard servings in hostelries from Galway to the Black Sea, from Nordkap in

Norway to Malta in the Mediterranean – but of shared heroes. There are nine of these, carefully composed in triplets some time in the fourteenth century and celebrated across western medieval Europe. They are three pagan, three Jewish and three Christian knights: Hector of Troy, Alexander the Great and Julius Caesar; Joshua who brought down the walls of Jericho, King David and Judas Maccabeus; Charlemagne, founder of the Holy Roman Empire, Godfrey of Bouillon who led the first Crusade in 1096 and ruled the fleeting Christian Kingdom of Jerusalem, and Arthur of Britain.

King Arthur was an early hero, and there are cadences in Malory's fifteenth-century version, first printed by William Caxton, that still thrill, especially the story of the sword in the stone ('Who so pulleth this sword from this stone and anvil is rightwise king born of all England'), and of the nobles who gather 'in the greatest church of London – whether it were Pauls or not the French book maketh no mention'. This reference to a French source is a reminder that Arthur is not exclusively British property. He pops up in Welsh and Breton poems, and he is listed as a heroic defender by Geoffrey of Monmouth in 1138. But versions of the Arthurian cycle also turn up in the same century in accounts by Chrétien de Troyes, who included Perceval or Parsifal and the Holy Grail, as well as the illicit affair between Lancelot and Guinevere; and Gottfried von Strassburg, who composed the story of Tristan, which also turns up in Malory.

Strasbourg is a city that – for me – symbolises the European ideal, and not just because it is host to a number of European institutions. It is French, but it is not like other French cities. It was, of course, at other points in its history, German. It changed hands five times in three generations: from the Franco-Prussian war of 1870 to the liberation of Europe in 1945. It is a city accustomed to thinking in French and German, and Alsatian too. It played host to Gutenberg when he was forced to flee from his home town of Mainz, and claims a place in the history of moveable type. It was once, famously, a free city, its own little republic. It plays a part – sometimes cruel, sometimes welcoming – in the history of the Jewish people in Europe. It seems an appropriate place to tell stories of fabulous people that somehow became the property of France, England and Germany too.

You can also see seemingly fresh and beautiful scenes from the stories of Arthur's court, and Tristan's passion, painted as murals on the walls of the Castel Ronco a few miles from Bolzano, in Italy. Bolzano railway station also announces itself as 'Bozen', and the castle too has a second name: it is the Schloss Runkelstein, because like Strasbourg in Alsace, Bolzano is part of a region with an identity crisis. The Italians call the district 'Trentino-Alto Adige', the Austrians call it the 'South Tyrol' or 'Sudtirol'. It had been part of the Austro-Hungarian Empire, it was annexed by Italy in 1919, after appalling bloodshed, and this was the cause of bitterness between native German- and Italian-speakers that continued until 1995, when Austria joined what is now the European Union, that entity referred to 'elliptically and allusively' as Europe in the only definition in the *Oxford English Dictionary*. This is the Europe first defined by the Treaty of Rome in 1957: this was the peaceful instrument that very slowly began to reassemble, uncertainly and with great bureaucratic awkwardness, first of all the Holy Roman Empire of Charlemagne, taking in France, West Germany and Italy, Belgium, Netherlands and Luxembourg, and then extending its embrace in one direction beyond Hadrian's Wall to Scotland and old Hibernia, Ireland, and Spain and Portugal on the Iberian peninsula; in the other direction to reassemble the Austria-Hungary of the Hapsburgs, and reach all the way to the shores of the Black Sea, and those countries that were part of ancient Byzantium, among them Greece, Bulgaria and Romania.

Unexpectedly, the Treaty of Rome has reconstituted most of the ancient Roman Empire, and added Poland and Hungary, and the forests and open plains from which the Goths and Visigoths first emerged. And this Europe had done these things not by force but by consent, by agreement. In my lifetime – I was born just in time to claim existence through most of the Second World War – the number of European dead in civil and military conflict numbers not millions but tens of millions: the numbers multiply according to your definition of the extent of Europe. But this was just the bloodshed, most of it accomplished before I started school, of one trifling lifetime. Two thousand years of recorded suffering – of plague, ruthless massacre, mass starvation, rape, slavery, torture, execution by burning, pressing, impalement, decapitation, defenestration, persecution, mutilation, remorseless slaughter on the battlefield, terrorism by pillaging invaders, concentration camps and extermination programmes, all of it brought to an end by a handful of peaceful treaties that began with the management of a continent's coal and steel, and the sharing of peaceful research into atomic energy. Other nations have been knocking on the door and asking for membership cards. Europe could one day embrace all that Rome ever ruled, and more, from the Atlantic to the Euphrates: one huge, free, assembly of states collectively managed by negotiated consent. The lesson of history is that no political structure survives for long. But for the moment Europe constitutes a continent, and a separate, entirely justified, line of an address.

SEVEN

The Hemisphere

A Divided World

The hemisphere is an almost meaningless form of geographical address that nevertheless puts me in my place, along with a great deal of unsought furnishing.

For forty years, a hemisphere label offered a badge of presumed loyalty – that is, we were expected to take sides – and a different hemisphere divide still implies membership of either the developed or the developing world. Divisions such as the Western world and the Eastern bloc imposed a cruel shape on twentieth-century history. Separations such as North and South still describe a divide between the haves and the have-nots. The paradox is that these notional geopolitical hemispheres have tenuous geographical substance, and certainly have never divided the world equally. But they represent powerful realities that imposed, and in some cases continue to impose, restrictions on movement, political activity, intellectual argument, publication, personal wealth and, above all, hope. If you drew them on a Mercator projection of the planet, or on a globe, or as a graphic representation of human numbers, or even of the distribution of human wealth, they would not look like half-worlds.

There is an infinite number of ways to divide a sphere into halves or hemispheres. A planet, however close it comes to being a perfect sphere, is quite different. This planet, along with all others in the solar system, can be – and is – notionally divided into equal halves, with boundaries that can be marked with exquisite accuracy; paradoxically, there are four of these halves. Because the Earth spins on one axis on its 365-day journey around the Sun, one of these divisions has a reality that would have been apparent under any circumstances: political, climatic, economic or geometric. Observers from an alien spacecraft would identify it immediately, and the first global explorers recognised its significance the moment they crossed this line. It is the Equator. Because the planet spins on an axis, there are two poles. Because there are two fixed poles – each representing a point with only one dimension, from which the horizon curves away in ever-increasing circles – there is an equal distance from each to a widest part of the planet, its maximum circumference. At this Equator, conditions exist which apply nowhere else. Because the planet spins at a tilt on its journey around the Sun, for six months of the year one hemisphere is more exposed to the Sun and gets more sunlight than the other. Days are longer, vegetation grows more energetically, average temperatures are higher. At the Equator, however, there are, in principle although not always in practice, no seasons, and a reliable twelve hours of sunlight every day of the year. This means that at any time, the Equator is a natural boundary between two zones. In one of these zones, the daylight lasts for longer, and the temperatures are likely to be higher, than in the other. In the northern hemisphere, the Sun at its highest point in the sky is most often to be in the south. In the southern hemisphere, the noonday sun is more likely to be in the north. The word 'likely' is necessary because there is an equatorial band – called the tropics – periodically traversed by the Sun. At the Equator, twice a year, in the March and September eqinoxes, the Sun is overhead at midday; once a year it is directly overhead at the Tropic of Cancer in the northern hemisphere, at the summer solstice in June; and once a year, during the winter solstice in December, it is directly over the Tropic of Capricorn in the southern hemisphere.

Nevertheless, an explorer who had no idea of calendar, and no understanding of planetary dynamics, would still soon realise that some strange, defining line had been crossed at the Equator. Average temperatures, having continued to rise on the journey south, would begin to fall. Prevailing winds would seem to come from a different direction. Skies would become clearer, and sunlight more intense, because the area of habitable land, and the density of human population, south of the Equator are so much lower, so that there would be a measurable fall in atmospheric pollution by dust, soot and tiny aerosols of sulphurous compounds. It would not be a journey into a looking-glass world, one of overall symmetry. The Sun would still rise in the east and set in the west, but as you sailed south it would increasingly be at your back at midday, rather than in your face. There would be even more dramatic evidence: new stars and unfamiliar constellations would begin to reveal themselves at night.

So the Equator marks an invisible but palpable divide: a great circle for navigators – a great circle that is the circumference of an imaginary disc that passes through the centre of the Earth, and the Equator is the only line of latitude that is a great circle – and an exact rather than a relative boundary between north and south. North is a real idea, with defining characteristics, and although the nearer you get to north, the more confusing the polar identity gets (you can have geographic north, magnetic north, geomagnetic north, the celestial north pole, the pole of cold, and the pole of inaccessibility, all points that mark a planetary extreme), there is something final about the geographic north pole. North stops at the north pole: there is no further north you can go. From the centre of the Arctic Circle, everywhere is south. The reverse holds true in Antarctica. The Equator had to be discovered, but it did not have to be invented. Someone born in New Zealand is a creature of the southern hemisphere. Someone who migrates from the South Pacific to the United Kingdom must enter and settle in the northern hemisphere.

The longitudes that define east and west, however, exist only as ideas that had to be proposed, and then measured with great difficulty but increasing precision by generations of navigators, and finally ratified by formal international agreement. That is because the Earth spins on its axis, delivering midday at one place, midnight at another. While families in Derbyshire or Sussex sit down to afternoon tea, people in Chicago might still be lingering over breakfast. As farmers in New Zealand rise to milk the dairy herd, Parisians have begun to think about an aperitif before dinner.

Because the hours chase each other around the spinning globe, because when it is day in one place it must be night at another, there must be a place where one day, too, succeeds another. When Jules Verne's Phileas Fogg travelled eastwards *Around the World in Eighty Days*, he assumed he had lost his wager, because he had marked a total of eighty-one days off his calendar, while those he left behind in London counted only eighty elapsed days. Had he travelled in the other direction, and triumphantly ticked off eighty days, he would have discovered that London's clubmen had counted eighty-one days, and had already pocketed their winnings.

This 'circumnavigator's paradox' first began to puzzle geographers at the beginning of the fourteenth century. Magellan's great traverse of the globe began in 1519. When one of the survivors of the voyage reached the Cape Verde islands three years later he was convinced the date was Wednesday, 9 July 1522. The sailors who went ashore for food and fresh water asked what day it was. 'They were answered in Portuguese that it was Thursday, at which they were much amazed, for to us it was Wednesday, and we knew not how we had fallen into this error,' wrote Antonio Pigafetta. 'For every day I, being always in health, had written down each day without any intermission. But as we were told since, there had been no mistake, for we had always made our voyage westward and had returned to the same place of departure as the sun, wherefore the long voyage had brought the gain of 24 hours, as is clearly seen.'

So to travel somewhere else is an adventure in relativity and space time. Quite soon it became imperative to limit the amount of adventure and settle on some exactness. Explorers needed independent evidence as to where on Earth they might be, and how they might find the same place a second time. They could gauge how far north or south they had sailed by fixing the noon-day altitude of the Sun and then consulting an almanac that would predict solar height according to calendar date and latitude. Longitude could in theory also be fixed by astronomical reference, but not very precisely, and not at all without some fixed reference point and some reliable technology. There are 360 degrees in a circle, and twenty-four hours in a day. So the Sun would inevitably progress through fifteen degrees each hour. If a mariner could carry a clock that marked the diurnal hours at his starting point, and if he could accurately fix the moment of midday at various points along his voyage of exploration, then he would also know how far he was from home, and precisely where on the globe he might now be.

In the modern world, travellers simply reset their wristwatches according to instruction from the locals, and keep in step with local time, but that doesn't precisely measure their longitude. Fifteen degrees, a few miles from the pole, is not a large span. At the Equator, however, it represents almost 1,667 kilometres, or a little more than a thousand miles. Nor does punctilious wristwatch adjustment settle the puzzling question (puzzling for a traveller) of the precise point at which to keep in step with people at home: to do this, global circumnavigators either experience the same day twice, or skip a day altogether, according to their direction of travel. Seafaring, trading and colonising nations needed to know such things, and maintained a number of initial reference points for centuries.

In October 1884, an international conference settled the matter once and for all – consider, for a moment, how precise that phrase is – by agreeing that universal time should tick away precisely in step with Greenwich Mean Time, and that longitude zero should pass through the Greenwich Meridian. Since the Greenwich Meridian was henceforward zero, to everybody on the planet, the International Date Line would then necessarily be at longitude 180 degrees, which fortuitously cuts through the empty Pacific Ocean for most of its journey. Someone who flies eastward from New Zealand on a Monday and crosses this dateline travels back in time to Sunday. This convention presents few problems for New Zealanders (except for the matter of the date on Chatham Island, a NZ dependency on the other side of the 180-degree meridian), but it would be a serious embarrassment in eastern Siberia, where it would be possible to cross back in time while sauntering to the next settlement. So the International Date Line comes with a series of agreed kinks in it, wiggling here and there to keep the last outposts of Siberia in the last Russian time zone, some Polynesian islands in temporal step with Hawaii, and Chatham Island on the same local time zone as NZ. Why not? These things are political conventions that serve largely economic and administrative conveniences. Local time in Bristol in Somerset is quite different from local time at Greenwich in London, but it suits schools, businesses, broadcasters and the managers of train and bus timetables to act as if it is identical. The continental United States has three time zones. Australia has three time zones, but paradoxically the northern states remain out of step with their neighbours directly to the south, on essentially the same line of longitude. Russia, extended across almost half of the northern hemisphere, has eleven time zones, but its railway operates only on Moscow time. China, which in 1912 had five time zones, is now content with one. So while the northern and southern hemispheres may share the hour and the day, but not the season, the peoples of the eastern and western hemispheres can adjust their clocks as they like for an hour or two in any direction, while overall remaining prisoners of night and day.

This geographical happenstance presents no problems at all if you know the rules that apply in each nation. But in the late twentieth century, history presented a different set of choices: between the West, or the Eastern Bloc, or the Third or non-aligned world, and these were, in effect, no choices at all. In February 1945, near the close of the Second World War, the leaders of three Allied powers met at Yalta in the Crimea on longitude 34 degrees East and effectively divided the world into Eastern and Western spheres of influence, and imposed a set of additional geographies, complete with amended historical baggage, politically organised loyalties and an uneasy and very imperfect peace that came to be based on the threat of direct mutual annihilation if direct mutual hostilities ever reached flashpoint. For many people, this peace was largely illusory, and for the next forty years East and West fought each other by proxy, by shows of force, by propping up unpopular regimes, by coups and military takeovers, or by encouraging or lending military support in a series of civil and border wars in Korea, what had been French Indo-China, in the Middle East, Iran and Afghanistan in former colonial Africa, in the Caribbean and Central and South America.

The notion of 'east' and 'west' as mutually opposed was not a new one: European nations had been nervous about the intentions of imperial and later, revolutionary, Russia, and then

China and Japan, for more than a century beforehand, and this distrust was reciprocated, to wretched and bloody effect. This notion of a divided world, a planet of two halves, even had distant roots in religious and political history: ancient Greece was obsessed with its Persian enemies to the east; Rome divided into a western and an eastern empire, and then a Catholic and an Orthodox Church. In 1453 ancient Byzantium, the Rome of the east, finally fell to the Ottoman Empire, and once again two great forces contended for control, or shared an uneasy and shifting frontier. But even in the farthest west of Europe, there was pressure to divide the globe. In 1493, a papal bull shared the world between two great imperial powers: Spain and Portugal. In an encyclical called *Inter Caetera Divinae*, Pope Alexander VI proposed an imaginary line from pole to pole, one hundred leagues west of the Azores and the Cape Verde islands. Discoveries west of this line were to be granted to Spain; east to Portugal. Brazilians, indeed, still speak Portuguese, but from the Rio Grande border of Texas to Tierra del Fuego, the other peoples of North, Central and South America mostly get by in Spanish. The lesson of this act of political piety is that it is not enough to issue a decree: you must occupy the ground and hold it, and the world that Alexander VI so airily handed to two Catholic monarchs was far bigger than many people supposed, had not been so far mapped by navigators from either nation, and was anyway to be seized in part by the English, the Dutch and the French as well. The world in 1493 was a vast place, and no single power held any part of it in a very secure grip.

But the imposition in 1946 of the Iron Curtain in Europe – a barrier that ran from the Baltic to Trieste – and the establishment of military bases within the Arctic Circle to forestall sudden aerial attack across the polar ice created two zones of fear and suspicion in what had become an unexpectedly small world: small in the sense that in October 1957, the Soviet rocket engineers who devised Sputnik 1 had found a way of circling the entire planet in ninety-two minutes; small in the sense that in the event of a surprise attack, people in Britain and western Europe could expect a warning of no more than four minutes before thermonuclear obliteration by missiles launched from behind the Iron Curtain. But this fear and suspicion had been founded on more old-fashioned ideological principles, on infiltration from beyond and betrayal from within, and on the more obvious fear of invasion, on both sides, by tanks and propeller-driven or jet-powered bombers, followed by men with bayonets. Such anxieties – driven by deliberate and sustained propaganda, once again from both sides – framed a mental and spiritual landscape that responded to the thoughtful novels of the period like Orwell's *Nineteen Eighty-Four* and *Animal Farm*, but also to other stimuli: the visual marvels and polemical pieties of Soviet cinema, and unconsciously or frankly propagandising US movies from the McCarthy period such as *Invasion of the Body Snatchers* and *I was a Communist for the FBI*.

I saw this last movie screened from a flickering projector at the Devonport fire brigade station some time in the first half of the 1950s, and I watched it already knowing – because that was the sort of thing you were told at Catholic schools in the 1950s – that the Soviet Communists were so evil that they encouraged children to inform on and denounce their parents. The story involves a tortured individual who is rejected by his own son because he is a member of the Communist Party, and remains rejected until that glorious moment of tearful apotheosis when he steps into the witness box and reveals that he had infiltrated the party as an undercover agent for the Federal Bureau of Investigation, and that he was therefore all along a true patriot. The implicit lesson – that furtiveness, subterfuge, betrayal and family rejection were OK as long as they were all in the cause of free-market capitalism, and that the American notion of liberty did not include the freedom to be a communist – may not have had its intended effect on me. But the film left me clearly understanding that there were two hemispheres, and that I was part of the Western world.

This, geographically, was ridiculous: New Zealand was part of the eastern hemisphere, along with the Japanese and the South Koreans, and those then-committed Western allies the Turks, the Pakistanis and the Iranians. I remained part of this stubborn eastern outpost of the Western world until late January 1961, when I crossed the Greenwich meridian, somewhere south of Andalusia, on the *Castel Felice*, bound for Southampton. I lived in London west of the Greenwich meridian, and therefore unequivocally part of the Western hemisphere, but only just. This marginal existence – with an address of zero degrees, so many minutes west of the meridian – continued in London and in the East Riding of Yorkshire until 1965, when my young family moved to Shepherdswell, in Kent, back in the eastern hemisphere, and stayed there until 1970. We crossed the Greenwich meridian again with a move to west London in 1970, and crossed back to the eastern bloc in 1985, with a move to Hastings.

It was during these years that the world was warned about a second geopolitical division, this time along an imaginary divide called the Brandt line. It took its name from a report by a commission led by the then German chancellor Willy Brandt. His commission pointed to a division roughly along latitude 30 North, north of which nations were economically powerful and their citizens were in most senses healthy and wealthy; south of which governments were weak or even unstable, and their peoples at risk from disease and hunger, and condemned to economic insecurity or crushing poverty. Once again, places like Australia and New Zealand were anomalies: outposts of the north, far to the south, their citizens increasingly aware of their good fortune, and increasingly nervous and guilty guardians of their own comfort. These placings on a series of imaginary lines – north and south, east and west, have and have-not, communist and capitalist, freedom-fighter and freedom-lover, aligned and non-aligned, Cyrillic and Roman, Anglophone and Francophone, English and Hispanic, Protestant and Catholic, Catholic and Orthodox, Jewish and Christian, Christian and Muslim, Muslim and Jewish – tend to pinpoint us, to leave us with a sense that we occupy a position, lock us somewhere on a grid of unsought loyalties and unspoken assumptions, that vary according to our national history, our family origins, our neighbourhood, our educational tradition, and our generation.

But such imaginary, invisible but somehow inescapable divisions of land and sea, the one-dimensional lines of latitude and longitude, have also provided us with something much surer and more precise. One of them, the Tropic of Cancer, told Eratosthenes, a Greek born in Cyrene in North Africa, later a librarian in Alexandria, that the world was round, and probably about 24,000 miles in circumference. The story – I know it only from popular histories – is one of the great legends of science. Supposedly in around 240 BC Eratosthenes was in Syene, in upper Egypt, now covered by the Aswan dam, during the summer solstice, and he observed that at noon the Sun was directly over his head, and he cast no shadow. In Alexandria, on the same day in another year, the Sun was at a slight angle at its highest point, so that he cast a shadow. He measured the angle of the Sun at Alexandria, and then calculated the distance due southwards to Syene, did a bit of trigonometry, and announced that one degree of the Earth's circumference would be seven hundred stadia, and therefore the Earth was 252,000 stadia in circumference. The stadium was a measure used by both the Greeks and the Egyptians, but the Egyptian stadium was a lot shorter than the Greek one. Nobody now knows whether Eratosthenes was doing his calculations according to the Greek standard of 185 metres, or the Egyptian one of about 157.5 metres. If he was thinking in terms of the Greek unit of measurement, he overestimated the size of the Earth. If he calculated in the Egyptian measure, he was eerily accurate: he put the Earth's circumference at more than 39,000 kilometres, within 1 per cent of the real figure.

There is a lot more to the story of Eratosthenes, but the lesson that matters most is the value of a permanent, frequently checked and internationally agreed standard for length – along with similar standards for mass and time. We have them now: the metre, the kilogram and the second. And all three of them owe their present precision to measurements of latitude and longitude. The third of these was the first to be established: the Babylonians divided the day into twenty-four hours. The idea of a day – a complete revolution of the Earth, marked by the Sun's position – was universal, but cultures, states and even cities could decide for themselves when this day began and ended: at midnight, sunrise, noon or sunset. The Egyptians, and the Romans followed the twenty-four-hour lead, but divided the day into twelve hours of night, twelve of day. Since day-length varies according to season everywhere except at the Equator, this left the notion of one hour as an elastic quantity, fine for making appointments with people next door, tricky for anything else.

This hour could be divided by sixty minutes – using sexagesimal arithmetic, first devised by the Babylonians – but the accurate division of time into anything much shorter than an hour was, for much of human endeavour and for most of human history, pointless. In fact, medieval monks anxious to time their daily offices as accurately as possible divided their hours (*horae*) into five points (*puncta*); from this, we get the word punctual. The idea of a smaller but precisely measurable practical unit of time had to wait for the arrival of science – Galileo experimented with a pendulum, and with his pulse – and the navigator's chronometer in the eighteenth century. The second remained a division of the day, and therefore the year, both of which could be measured from any meridian or fixed line of longitude. A day was counted as 86,400 seconds.

By 1956, however, geophysicists and astronomers had begun to realise that the Earth wasn't itself a reliable piece of clockwork, so an international conference fixed the second as a very precise division of the tropical year. A tropical year is what you would see from a particular meridian, as the Sun returns through the cycle of seasons to exactly the same point. So, rather than a division of a minute, an hour or a day, a second became a precise fraction of a year. But since a procession of years might vary each from the other by a fraction of a second, astronomers settled on one particular reference year – 1900 – as the standard by which to define the second. This was, in every sense, a temporary measure. Shortly afterwards, with the invention of atomic clocks, the length of a second was fixed as so many oscillations of a caesium atom at a particular temperature, and the second no longer had anything to do with meridians of longitude. But it remained a notional measure of the day, from a line that would divide the globe into eastern and western hemispheres.

The metre had a similar origin, and a similar end. It began as a systematic attempt to take the measure of a meridian. The word metre itself stems from the Greek for 'measure'. European nations made do with miles nautical and statute, knots, leagues, cubits, stadia and versts, with fathoms, rods, poles and perches, furlongs, chains, feet, yards and so on. In 1790 French authorities decided on a universal unit of measurement that would be termed a metre, and the original proposal was that it should be based on the distance back and forth covered by a pendulum within two seconds. But experiments showed that pendulums behaved differently in different parts of the country, so the French Academy placed the problem before a scientific commission which in 1791 suggested that the new unit of measurement should be one ten millionth of the distance from the North Pole to the Equator, at sea level. Nobody had at that time been to the North Pole. Nobody knew how to measure a straight line across the Mediterranean and the Sahara accurately all the way to the Equator. Nor could anybody be sure whether the planet was – underneath all its ups and downs, its bumps, crevices, soaring volcanoes and eroding mountain folds – a true sphere (in fact it is an oblate spheroid, slightly flattened at the poles, and with a little bulge at the Equator). Two French astronomers began a survey of the meridian that runs from Dunkirk in France to Barcelona in Spain. In effect, they would do what Eratosthenes did: measure a little bit of curvature and extrapolate from that. In practice, and unlike Eratosthenes, they would do it painstakingly and in fine detail, covering every footstep along the way, and spend years on the job if necessary. One of them, Jean-Baptiste Delambre, took on the relatively flat terrain between the English Channel and Rodez in southern France, the other, Pierre François-André Méchain, started from Barcelona and worked his way across the Pyrenees towards Rodez.

The surveyors kept count in an old French unit of measurement called a *toise*, but even this measure explains why a standard, universal measure or metre might be a good idea. The *toise* traditionally used for surveying land in the Cantal region was considerably longer than the one customary in Strasbourg, but shorter than the *toise* used in Besançon, so the surveyors started with an agreed measure called the *toise du Perou*, based on an iron bar that had been used by a French team for a survey of

Ecuador. The project had begun at the suggestion of the statesman Talleyrand, Bishop of Autun, and had been authorised by Louis XVI. It survived the French Revolution, the execution of the king and queen, the Terror, the abolition of the Academy of Sciences and war with Great Britain, Spain, Austria, Prussia and the Netherlands.

At the end of the project, the surveyors proposed a length for the metre. In June 1799 a jeweller produced a thin, flat ruler of platinum that became the Mètre des Archives, and thereafter the definitive metre. This standard measure was not, in fact, quite one ten millionth of the distance from the Equator to the pole – satellite measurements from space can achieve accuracies that no eighteenth-century explorer could have dreamed of – but it was pretty close all the same, and in a series of international conferences it was adopted formally by the rest of the world.

The unit of mass, too, was a revolutionary decision: in 1795, the French decided that this would be called a gram, and defined as the mass of water, at the temperature of melting ice, contained in a cube the sides of which measured one hundredth of a metre. Since nobody purchased cloth, sold bread or cast metal in measures as small as a gram, the unit quickly became the much more convenient kilogram.

Two of these measures have now been detached from their original connection with meridians and the Equator. As we've seen, the second is now so many oscillations of an isotope of caesium. The metre is the path travelled by a precise frequency of laser light in an unimaginably small fraction of a second. At the time of writing, the kilogram is beginning to cause worry among those experts paid to worry about such things. That is because the international prototype kilogram stored in a vault at Sèvres near Paris seems to weigh micrograms less than its official copies distributed through other countries, so one day the kilogram, too, may be entirely abstracted from its Earthly origins, to be defined on the basis of something that could never oxidise, abrade, erode or even adsorb extra molecules from the environment; as something that would always be the universally agreed test for a kilogram.

Huge scientific prizes hinge on the ability to time, measure and weigh things with very great accuracy. Vast international scientific endeavours – gravity-wave detectors that might one day float far away in space, huge particle accelerators already buried below France and Switzerland – operate on the assumption that time can be measured in billion billionths of a second, that particles that exist for periods counted in these attoseconds can still be assigned a mass, or the energy equivalent of mass, and measured over distances smaller than the radius of a hydrogen atom. With the knowledge acquired by instruments as huge, and as sensitive, as such detectors, cosmologists speak with confidence about the history of the universe in its first billionth of a second, when the cosmos might have been the size of a basketball. So the second is now a universal measure: as enduring or as fleeting on the surface of Titan or near the heart of Andromeda as it is at the Greenwich or the Paris meridian; and the metre will plot the ice flows on Europa or the circumference of Betelgeuse as reliably as it will measure a bolt of cloth at a draper's shop in east London. From the kilogram, the metre and the second, so many certainties follow. Without them, there would be no universal standard measurement for density, angular velocity, or heat capacity; for volume, area, acceleration of volumetric flow; for chemical, gravitational or electrical energy. These all derive from the second and the metre, and both of these originally derive in turn from calculations that involve a line of longitude, any one of which divides the planet into eastern and western hemispheres; and the Equator, which divides the planet into a northern and southern hemisphere. Ultimately, everything about me – glucose use in the cerebral cortex, for instance, or the mass of epidermis shed during a summer's day, or the blood pressure in my upper arm – can be described in terms that owe their origins to the arbitrary division of the sphere into meridian lines that girdle the planet and divide it into hemispheres.

EIGHT

The Planet

Down to Earth

We live in a makeshift home with a limited tenancy. It was more than a hundred million years in the making, and it is due to be destroyed by fire in a few billion years, baked to a crisp by its parent star; but life's leasehold will expire long before that, along with the planet's atmosphere and water. *Gaia* to the Greeks, *Terra* or *Tellus* to the Romans, the only known domicile for life is an unstable, not-exactly spherical arrangement of metal, mineral, gas and water with a molten core, a viscous mantle and a hard but shifting crust. It is for the moment 29 per cent rocky surface and 71 per cent salt water, but these proportions are subject to small changes over short time scales, and quite big changes if you set the ticking of the clock to intervals of many millions of years.

If you were looking for the ideal place to settle, you might not begin with Earth as your first choice, but life is a matter of compromise, and in any case a celestial estate agent might persuade you that the planet has a number of significant advantages. It has hot and cold running water already laid on. It is air conditioned. It has a fully working invisible deflector shield far beyond the atmosphere that protects the surface from episodic solar artillery bombardment, and – this is a bonus – this same screen broadcasts a periodic random plasma display of ethereal beauty to brighten the polar nights. Early tenants unwittingly installed a second, fragile, invisible screen high in the strato-sphere to protect all subsequent leaseholders from lethal ultra-violet radiation. This planet comes with reliable long-term provision of electromagnetic energy at an average rate of 1369 watts per square metre. That is quite enough to power a micro-wave oven, a hair dryer or a toaster, not to mention almost all life on Earth.

Besides a steady power supply the planet is also equipped with an ethereal form of loft insulation, and heavy-duty under-floor central heating that permits no inspection and requires no repairs. It offers a range of habitats to suit all tastes. Weathering and mechanical attrition are continuous, and intermittently very troubling, but this global household wear and tear is balanced by a convenient autonomous restructuring process that builds up new mountains as fast as older ones can be worn down, and tears open new fjords and harbours as fast as old ones become silted up. This same process recycles water and atmosphere, delivers freshly minted mineral nutrients on an almost daily basis and slowly converts old bones and worn sedi-ment into marble and slate. This is a planet capable not just of encouraging life, but of sustaining it on a global scale, all the while laying down conveniently accessible energy reserves and pools of raw material for bricks and mortar. Put like that, Earth sounds a bargain: no down payment, no mortgage, no charge for solar-powered electricity; all the materials not just for survival, but even for occasional comfort; neither too hot nor too cold; in fact, just right.

But why is it just right? Will it continue to be just right? Was it always just right? Are we here at life's lucky moment? This is what planetary scientists call the Goldilocks problem, the lesson in thermodynamics at the heart of the fairy tale of the little girl, three bears and three different-sized plates of porridge, one of which is too cold, one too hot, the other just right. It doesn't take much more than rudimentary physics and a perfunctory experience of the breakfast table to realise that if all the bowls of porridge were poured from the saucepan at the same time, then the deciding factor would be the time of arrival of Goldilocks. If she had turned up a bit later, the small and middle-sized bowls would have been too cold, and Father Bear's porridge would have been just right. If she had arrived too early, all three bowls would have been too hot. The logic is that, at different points in the Goldilocks continuum, any one of the bowls would have been just right.

What is true for porridge is true for planets: planet Earth, too, would have arrived smoking hot. What right now looks wonderfully like a best-in-show at a notional Ideal Planet Exhibition was, like some cosmic cowboy construction, literally thrown together. The polite or scientific way of saying this is that over a period of thirty to forty million years, the planet 'accreted' from planetismals, or dismal little planets. These planetismals would have been lumps of metal, mineral and miscellaneous dust about ten kilometres across. Such objects – one thinks of them as bigger than Manhattan, or Hastings and St Leonards together – were themselves accretions, little assemblages of gas, dust and gravel that condensed from the disc of stuff swirling around the newly formed star called the Sun. What began as wisps of matter, dancing in the gravitational embrace of the newborn star ninety million miles away, began to gather substance, momentum and mass, and to form into loose aggregations of dust and small stones. In the freezing emptiness of space, these would nonetheless have warmth. They would start to get hot from the friction of contact with each other, and some of them would provide their own inner warmth, because they would have been fashioned not just from metals but from radioisotopes of metals: atoms of aluminium or iron with more than the ideal number of particles in the nucleus. That is, these isotopes would be unstable. They might last for minutes, they might last for a hundred thousand years, or a hundred million, but sooner or later they would decay, and send an electron or a proton or a neutron whizzing away, releasing a burst of heat. Enough of them, and these decaying isotopes could turn their parent planetismal into a hot fudge of molten aluminium, iron, tungsten, nickel and anything else. The bigger the planetismal, the hotter it would become. These objects would be tearing around the Sun at thirty kilometres a second, bashing into each other, rebounding, slowing down or speeding up, ricocheting away and slamming into something else. Any objects that collided at the right angle or the right speed would merge, become one: after a while, their accumulating gravitational muscle – the force that began the accretion of dust and gas molecules in the first place – would begin to make a big difference: smaller planetismals overtaken by the big ones would become rolled up in them, subsumed. They would surrender their identity, but maximise their staying power. You could make a mantra of it: the planetismals that prey together, stay together. The bigger protoplanets would go on sweeping up the smaller ones, and the bigger they were, the longer they would retain their heat. That is because the ratio of volume (the whole collection of hot minerals) to the surface area (the membrane through which the heat leaks into the freezing outer darkness of space) is bigger. Size matters. It pays to be substantial.

If a protoplanet gets big enough and hot enough, something of enduring significance begins to happen. The mix of hot, heavy metals begins to separate from the bubbling gruel of other minerals, such as silica, and sink to the centre. Once far from the surface, it takes longer to cool. It is insulated from the outside, for a start. And then it still contains its own mix of decaying isotopes, which emit heat that can then only escape very slowly, because of the insulating layer of other things around the hot ball of molten iron and nickel at the centre. The outside, too, would be molten and smoking hot, because other objects would continue to strike it, and strike sparks from it. The planetismal doesn't have to be very big for this orderly meltdown, this separation of heavy metals from lighter metallic and mineral compounds, to happen: about 10 per cent of the size of the Earth. But once it begins to happen, there is a chance of becoming a Goldilocks planet. The object won't be the only heavy-duty survivor: the constant bashing and crashing around the solar racetrack must have begun with hundreds of planets-that-might-have-been. This violent, careering joyride of competitor protoplanets ended with Mercury, Venus, Earth, the Moon, Mars and its two little moons, and the asteroid belt. Each of these is the product of countless collisions, course corrections and speed adjustments that settled down into an orderly and more or less predictable procession around the Sun.

There is both a price to pay and a reward to reap for removing the competition. The survivor sweeps up most of the dust and small stones around it, and becomes – in the course of a hundred million years or so – a less dangerous place: an outer layer of hot or molten minerals coating an inner core of molten iron and nickel. Too hot for comfort, far too hot for life, and with no guarantee that it won't smack into something else really large and sharing the same orbit; there is a long way to go and a lot of settling down and many nasty surprises ahead, but it would represent all the makings of a desirable property, a place in the sun. It would be surfaced with oxygen, silicon, aluminium, iron, calcium, sodium, potassium, magnesium, hydrogen and many others, but these atoms would have already started doing deals with each other, forming alliances, forging close bonds: settling down into those mineral mixtures known as magma when molten, rocks when stiff and cold. But even as this outer layer began to cool and become rigid, it would still be gouged, hammered and caressed by steady deliveries of additional raw material from outer space: meteoroids rich in nickel and iron and carbon, thumping in at twenty kilometres a second; comets, strange loose aggregates of frozen snow and dirt and mysterious chemicals, landing at up to seventy kilometres a second.

The comets, in particular, play a powerful role in this story. These seem to serve as celestial laboratory benches: they sweep up the elements and compounds that might already exist in the space between the stars. Water is one of these compounds, but there are many others: alcohol, for instance, and formaldehyde, and hydrogen cyanide, benzene and acetic acid, sodium chloride and formic acid have all been identified drifting about the galaxy by astronomers. On the frozen surfaces of comets, these and other chemicals meet and mix and form complex arrangements that nobody could ever have predicted: hydrocarbons, for instance, and amino acids, the building blocks of protein. When somebody unravels the great story of where life came from, comets will figure in this narrative, as pre-proto-progenitors that delivered the stuff that would somehow become the seed and egg of life; or ghostly godparents that put down an endowment fund of ready-made wealth that life's first manifestation could exploit and build upon; or as life's first home-maker, because life began in the sea and comets almost certainly provided some though not all of the water that now washes two thirds of the planet: asteroids probably provided most of it, along with much of the planet's primary chemistry set.

But that is to anticipate: within the first hundred million years of its existence, the Earth-to-be was a hellish place (geologists call it the Hadean period), marked by repeated violent assault from above as great lumps of rock, metal and ice fell like rain from the hostile heavens. As the ball of stuff that was to become Earth got bigger and bigger, so did other contenders for the same space. One of these almost certainly slammed into Earth, to dislodge enough rock to form a partner now called the Moon, a catastrophic accident that, far from writing off Earth as a home for life, may have helped life's genesis and evolution. The Moon's stately gravitational dance around Earth keeps the planet's axis more or less stable, so that the Equator is always hotter than the poles, and stays that way. With stable temperature zones, life can not only get going, but start adapting, a little at a time. At the moment of the Moon's making, there could have been no place or prospect for life. But as the protoplanets got bigger, the orbits within the potential zone of life, the region near enough to the Sun for liquid water to exist, became less cluttered. There developed clear traffic lanes, reserved for Mercury, Venus, Earth, Mars and the asteroid belt, a stream of rubble widely thought of as a planet that somehow never accreted.

The buffeting continued (it continues to this day: 40,000 tons a year of stuff from the solar system flickers through the atmosphere in streaks of light, and some of it hits Earth's surface), but it became possible for the protoplanet to settle down and become a planet. It developed an atmosphere, and an ocean, and it began – very slowly at first – to develop a continental surface. It could do this because it was hot, and alive: the great ball of molten metal at its core, and the fissile uranium and other radioactive elements in its mantle, kept its interior warm and viscous, and set up a kind of geochemistry apparatus that began to send huge, slow-moving globules of molten granite towards the crust; that marshalled volcanic forces that began to spew molten lava, hot gases and superheated steam to the surface to cool as rock, to linger as air or to fall as rain.

This planet would still not be recognisable as Earth. It would have no America, no Asia, no Africa, virtually no land at all. Its sky would not have been blue, because there was no oxygen to scatter the sunlight and fabricate the illusion of an azure vault. Its first ocean would not have been saline, because the first rains would have barely begun to dissolve salts from the first rocks. Its day would certainly not have been twenty-four hours long: the early Earth spun on its axis much more swiftly. The ball of molten metal at the Earth's core, however, would already have begun to serve as a dynamo that would excite a magnetic field which would extend far beyond the upper atmosphere, and double as a *Star Trek*-style deflector shield, catching, trapping and disarming the death-dealing material hurled from the erupting Sun. Once this deflector was in place, life stood a chance. The first fragile biochemical attempts at self-replication might begin at any time in the protection of the first seas, but would not have survived for long at the surface, under intermittent but inexorable blasts of buckshot from the Sun.

The Earth's new companion, too, would have begun to exert a force, creating tides not just in the planet's first seas but even in its rocks that would, eventually, slow the Earth's rotation to its present twenty-four-hour day. Consider the timetable: the solar system has existed for about 4.5 billion years. So the Earth is notionally about 4.5 billion years old. Nobody can be sure how long it took the inner racetrack of the solar system to settle down into a set of rocky planets, but it must have happened within the first five hundred million years, because the evidence in the planet's surface rocks suggests that life first appeared on Earth around 3.8 billion years ago. Or perhaps that should read *last* appeared on Earth 3.8 billion years ago. Who knows how often the nascent habitat was hammered by huge comets, heavyweight planetismals and thumping great asteroids? A direct hit from a lump of space rock the size of Scotland, for instance, would generate enough heat to flash all the Earth's oceans into steam. It would vaporise unimaginable quantities of rock and send the flaming dust up into space and back again, encircling the globe and setting the planetary thermostat to getting on for 2,000°C, creating an atmosphere so dense you could swim in it, and of course sterilising the surface utterly. If such a life-obliterating thing happened once, it could have happened again and again. This suggests that the emergence of life from the original terrestrial biochemistry set was not the result of some fabulously improbable chain of circumstance, a wild throw of the dice in the intergalactic casino, but a happy accident waiting to happen, and happen again and again. It suggests that the cosmic dice might be loaded in life's favour. This in turn raises the other great question: if so, where are the neighbours?

Never mind how life began, and what the last surviving universal common ancestor looked like. (Don't even try to think of those strange forms that might have been taken by all the would-be creatures that didn't survive.) Its emergence, and its seeming restriction to just one place in the universe, is the big mystery, and this in turn is surrounded by a set of smaller mysteries. The most immediate of these is the Goldilocks question: why Earth? How could there have been liquid water on Earth four billion years ago? The Sun, too, is a dynamic object: it is a star that was born, grew to maturity, will have a lifespan appropriate to stars of its class, and will die in the manner of other such stars. So it too would have begun its stellar career in a shy, understated way: its radiation would have been much fainter four billion years ago, and the Earth would have been too cold for liquid water to swirl around on its surface. Most of the planet's surface, at that stage, would have been watery: the continents are a relatively late construction. So the early Earth would have moved from fire to ice. Once it was a ball of ice, the Sun's radiation would have bounced brilliantly off its surface – white reflects all wavelengths – and the planet would have stayed frozen for billions of years. Without liquid water, there could not be life as we know it. But we are here, so something must have happened to keep the water circulating and make the Earth a suitably mellow home for meerkats, melons and Methodist ministers. The best answer so far is a simple one: the atmosphere helped. There was enough carbon dioxide and water vapour – and perhaps methane too – to let the sunlight through and then trap the consequent infra-red radiation: in short, to convert the Sun's energy into practical, ice-melting warmth. The greenhouse effect may right now be creating little local difficulties for some of Earth's current leaseholders, but from the start it screened, nurtured and protected life on Earth; and, it begins to seem, only life on Earth.

Why, where, when and how life began on Earth are questions that remain not just open, but wide open. For about 1,800 years, most of Christian Europe was perfectly happy with the idea, derived from Jewish scriptures, that a divine providence fashioned all creatures separately and in one short effort, in an enclosed paradise or park, in or near what is now Iraq, about six thousand years ago. Confidence in the supposed six-thousand-year time span for creation evaporated as soon as the first geologists began to study not just rocks but the processes that seemed continuously to form and amend them. And the 'big bang' doctrine of life's emergence began to look less than explanatory as soon as the first zoologists and botanists – often also devout believers – began to study the intricacies, similarities and inter-dependencies of life around them. Charles Darwin, one of the first to develop the idea that all life's variety branched continuously from one original stem, proposed that life emerged in a warm little pond, possibly from some soup of pre-existing proteins and other organic chemistry. Later scientists have proposed that life could have first accidentally fashioned itself in the cold but still liquid ocean under an insulating shield of thick ice, or that it could somehow have happened around geothermal vents deep under the ocean. In 1977 biologists were astonished to find communities of animals – three hundred species at the last count – living around volcanic vents, scratching a living ultimately from superheated brines that gurgled up from deep in the Earth's crust, miles below the ocean surface, without benefit of sunlight.

The chemical soup from which life was first brewed isn't the problem. The stuff must have been available by the bucketful. Fifty years ago, US scientists blitzed things they thought must have existed in the primitive atmosphere – water, ammonia, methane and hydrogen – with an electric current and then analysed the results: at the end of the cookery session they found in their vessel a mix of organic chemicals that included amino acids, which are the bricks from which life builds proteins. A few years later, another experimenter found that a solution of ammonia and hydrogen cyanide was all it took to produce not only amino acids, but a chemical called adenine: this is one of the four building blocks of DNA, the code of life. So bolts of lightning might have helped simmer the haphazard ingredients from which the recipe for life ultimately emerged.

Some of this raw material for life would have already existed on Earth, independently of any electrical cooking process. It would have been delivered by meteorites. A lump of stone that fell from the skies over Murchison, Victoria, Australia, in 1969 turned out to hold around seventy amino acids, including many of those used in protein-building, as well as other organic chemicals such as the precursors of glucose, life's fuel. The discovery in 1996 of something that might – just might – have been a fossilised bacterium inside a meteorite known to have come from Mars raised another possibility: that life might have got started even earlier, on the Red Planet, and been transferred by accident, in the shrapnel from yet another explosive cosmic collision, from a place of origin that would once have been just right but soon became too cold, to an Earth that was now a much more likely home for life: warmer than Mars, but not too warm, cooler than Venus, but not too cool. For this story, it is enough that life began on Earth. It is here now, and so far it has been found nowhere else. It is fair to call life a miracle, because the word derives from the Latin *miraculum*, a thing to be wondered at. That is what the origin of life remains: a thing for wonder. We may never find out.

I am fond of thinking of the origin of life as a murder mystery in reverse. A body came to life in a locked library and released itself, before destroying the key and demolishing the library. Thereafter, all subsequent life forms trampled over the fragile evidence of this non-crime, probably destroying forever any hope of forensic evidence, and perhaps also making sure that such a creative deed could never happen again, leaving the planet occupied by myriad life forms all of which share only one lineage.*

For this story, what matters now is what Earth did to life, and what life did to its only known address. Life had acquired a home, a habitat, a place of its own, and proceeded to adapt to it, but also to amend and alter it to make it even more suitable for future generations. Life found a way to make a living, with no

---

*How inert bits of biochemical material somehow assembled into living, breathing, consuming, excreting and replicating organisms remains one of the great mysteries. The great British scientist Fred Hoyle in *The Intelligent Universe* (1983) likened the accidental emergence of complex life from organic raw materials to a tornado, blowing through an aircraft junkyard and suddenly assembling a functioning Boeing 747. Most biologists think the analogy is a poor one. Cesare Emiliani, in *The Scientific Companion* (1995), instead offers a twist on that old favourite, the monkey-typewriter-Bible challenge. There are six million characters in the text of the Bible, and twenty-six letters in the alphabet. The probability of one monkey, or even a billion monkeys, accidentally typing the Bible is effectively zero. But suppose, when a wild monkey randomly hits a key, the environment automatically erases that letter unless it is the right one. At one letter per second, and assuming thirteen errors per letter, the monkey will produce the Bible in thirteen times six million seconds, or thirty months. The environment, Emiliani argues, 'knows' what kind of organism would best fit. 'Given the chemical and environmental conditions of the primitive Earth, the appearance of life was a foregone conclusion. Only divine intervention could have kept Planet Earth sterile.'

further help from the great beyond, apart from a permanent supply of energy from the Sun. And it lived off the land, quite literally: all nourishment, all food, all living tissue, is a kind of conjuring trick: a confection whipped up from gases, minerals, a universal solvent and a reliable supply of radiant energy: in other words, earth, air, fire and water. The warm, self-renewing tissue of a living creature is the midpoint in a cycle that begins with the four classical elements and ultimately returns to them.*

---

*Vivid ideas never quite go away, even when they are superseded. Empedocles was born in Sicily around 490 BC. He wasn't the first to wonder about the basic elements of the universe, but he is supposed to have been the first to propose that all things were made from earth, air, fire and water in various combinations. He is also supposed to have thrown himself into the crater of Mount Etna in 430 BC. The idea that life collectively but unwittingly manipulates the planet's atmosphere and thus its temperature for its own benefit began more than three decades ago, as the Gaia hypothesis, and provoked many geologists and biologists into fits of near apoplexy. Its begetter, James Lovelock, has since been festooned with scientific honours, his proposal is now called the Gaia Theory, and in one modification or another it is widely accepted. He argued in *Gaia: A New Look at life on Earth* (1979) that if the Sun's surface temperature was 500°C, rather than 5,000°C, 'and the Earth were correspondingly closer, so that we received the same amount of warmth, life would never have got going. Life needs energy potent enough to sever chemical bonds: mere warmth is not enough.' So the fire element has to have firepower as well. That air is an element of life became apparent two centuries ago. Gabrielle Walker tells the story, in *An Ocean of Air* (2007), of Stephen Hales, an eighteenth-century clergyman who baked pigs' blood, deer horn, peas, tobacco, oil of cloves and gallstones, and derived air from all of them. He baked heart of oak, and found that it released 216 times its volume of air. 'Now 216 cubick inches of air, compressed into the space of one cubick inch, would, if it continued there in an elastick state, press against ... the six sides of the cube with a force equal to 19860 pounds, a force sufficient to rend the Oak with a vast explosion,' Hales wrote in 1727. He did not, and could not, know the precise chemistry of his unoaked gas, but it was certainly derived from the air in the first place. The Russian ecologist Vladimir Vernadsky in 1924, in an influential text called *La Géochimie*, later translated as *The Biosphere*, wrote that a certain huge swarm of locusts 'expressed in terms of chemical elements and metric tons, may be seen as analogical to a rock formation, or, more precisely, to a moving rock formation endowed with free energy'. This was simply another way of saying that however life began, it could only be fashioned from the inert elements around it. These, of course, exist in the air, and in the Earth. And water? Without liquid water, there is no life. A seventy-kilogram man is 62 per cent water by weight, meaning that he contains forty-three litres of water. He is a polymer sack filled with seventy-six pints of the stuff.

Yorkshire limestone and Sussex chalk are carbonate rocks made from Jurassic corals and Cretaceous coccolithophores. The carbon in carbon dioxide was once the trunk of a tree, the nitrogen that makes up four-fifths of the atmosphere was once incorporated in swiftly growing plants that were devoured by herbivorous animals which then discarded nitrogenous wastes that became the nitrate fertiliser that would encourage plant growth. When you see starlings in flight, you see so much stone and mineral moving across the sky; when you see a tree, you see tomorrow's air, or coal for some distant generation; when you see a decaying cadaver, or a rotting vegetable, you see a return to the atmosphere of warmth that was once radiant light. Life is a transient, fleeting partner that exists only in a dance of water, rock, air and sunlight: of earth, air, fire and water. These things existed before life; they will endure after all life has been erased. What has only recently become apparent is that, once life joined this elemental quadrille, it seemed to take charge, to set the pace, to exploit the innate rhythm that began as a purely mechanical process on a haphazard planet, and to maintain this process not just to sustain a community of interdependent planetary lease-holders, but to improve and enhance the property for future generations.

This is the point at which the Goldilocks metaphor begins to break down. Life has occupied the globe, adjusted the air conditioning, taken at least temporary charge of the thermostat and begun to impose a different and perhaps unique colour scheme. None of this happened in a hurry. For at least two billion years, the planet was controlled entirely by micro-organisms, creatures invisible to the naked eye, but numbered in their trillions in every handful of soil, scoop of mud or glass of water. These creatures are really all that matters: they are life on Earth. If life is a journey, then these creatures provide the hull and superstructure, navigational equipment, fuel, engine, steering gear, galley and quartermaster's stores as well. They manage the air, the water, the nutrients, the digestion and the recycling; they grow, cook and clean up after themselves. They turn carbon dioxide into carbon and fill the air with oxygen; they make the chlorophyll that harvests sunlight; and they turn the nitrogen in the air into nitrogen fertiliser in the ground. So they make the sky blue, the grass green and the soil rich. Microbes are survivors: they have been found on wind-blown dust high above the Atlantic, in water droplets on the surfaces of clouds, in pools of acid, lakes of natron, deep in sub-oceanic rocks, in the boiling water of geothermal vents, baked in the bricks of the pyramids, in the digestive tracts of mummified mammoths frozen for 10,000 years, and even on instruments sent by Apollo scientists to the Moon. They exist in numbers so enormous that words such as 'astronomical' have no real force. In a ton of soil, there could be a hundred million billion little microbes called prokaryotes. There are probably only a hundred billion stars in the galaxy. The microbial occupants upon and within the human body are supposed to outnumber tenfold the equally invisible cells in the human body. That means that each of us is little more than a thermostatically controlled, centrally heated, perambulating apartment block for a hundred trillion tiny tenants, and that some of these tenants must have arrived at our birth and yet more will be involved in helping dispose of us once we go into the eternal dark.

We know nothing at all about the first little creatures that emerged from the clumsy chemistry that began in the seas around the nascent Earth, but we do know a little about their descendants, because they left their mark on the young planet. Cyanobacteria, sometimes called blue-green algae, used sunlight, water and carbon to make their own food and form their own tissue. Since there was no sex – a late invention in life's story – and no hungry predators, and since death is an entirely notional condition for a creature that multiplies itself by subdivision, these tiny invisible things began to form thick mats of living tissue, now preserved in the planet's oldest fossil rocks, called stromatolites, laid down more than three billion years ago. Later than that, but more than two billion years ago, the same cyanobacteria evolved into oxygen producers and started releasing a lethal gas into the atmosphere, and other bacteria began to adapt to life with oxygen. Once that happened, the planet began to change. The skies became blue, because that's what oxygen does once there's enough of it: it scatters sunlight in the blue region of the spectrum. The surface also became safer, because some of the oxygen molecules split to become ozone, to absorb ultra-violet radiation and form a tenuous, fragile shield in the stratosphere, known as the ozone layer. So we have the first evidence that life had begun to take charge of its own habitat: it nourished itself from air and rocks and water and sunlight; it consumed some rocks, it made other rocks; it adjusted the atmosphere; it coloured the skies and it devised its own Sun-protection factor.

What happened next is still only guesswork. The story of the first few billion years is confused. Having been forged in circumstances so intense that lava must have foamed like water over its surface, did the Earth then become so cold that it began to resemble a snowball in space, leaving glacial scars all the way to the Equator? Did life suffer a series of shocks that all but wiped it out, again and again?* Did the planet's crust begin recycling itself as soon as it cooled enough to become a crust, or did it take another billion years or so for the mechanism of plate

---

*I once asked an asteroid expert what would actually happen, second by second, if a monster called asteroid 2002 NT7 slammed into the planet. This at the time seemed a not quite remote possibility. By the time he got the message, a deadline had passed. Nevertheless Dr Matt Genge, of Imperial College, promptly replied as follows:

tectonics to begin? Evidence exists for all these possibilities, but none of it is compelling. Whatever the answer, once huge wedges of the Earth's crust began banging against each other, or sliding under each other, driven by convection forces in the mantle far below the crust, the Earth began to look like an increasingly desirable residence. It had acquired blue skies, blue seas, pyrotechnic volcanoes, soaring mountains, variable winds and rainfall patterns and a mobile, slowly-changing terrain that would offer an astounding variety of habitats for the first plants, and

---

1. Asteroid 2002 NT7 passes the Moon. It will take about a day to reach the Earth's surface.
2. The asteroid reaches the top of the Earth's atmosphere at a speed in excess of twelve kilometres per second. Impact is in ten seconds. With its enormous size and speed, however, it starts to produce an enormous fireball that will be seen over hundreds of miles in every direction.
3. In a few seconds 2002 NT7 has penetrated deeper into the atmosphere, and the light and heat from the fireball become so intense that vapour begins to rise from the surface of the ocean below. There is no sound to warn of its approach; the asteroid is moving so fast that like the sonic boom of a supersonic plane the sound trails behind the fireball. The four-kilometre-wide asteroid has punched a hole in the atmosphere and is dragging outer space down towards the ground.
4. The asteroid, hidden in a white-hot fireball of heated atmosphere, ploughs into an ocean as if it wasn't there. The water at this speed is, however, like concrete. An immense shockwave tears downwards towards the sea floor, only just keeping ahead of the asteroid which, although rock, is deforming into a pancake. Much of the asteroid is now turning to gas due to the heat that is generated; much of the ocean at the target is turning to steam.
5. Microseconds later the asteroid reaches the ocean floor with the sea water flowing outwards to make a giant cavity tens of kilometres across. The collision of the asteroid with the muds of the sea floor hardly slows it at all and it continues downwards into the basalts of the oceanic crust.
6. Blistering hot gases and dust are expanding upwards out of the widening hole in the ocean in a giant plume. Some of these will be cast back out into space to fall back on the atmosphere and heat the Earth's surface surrounding the ocean. The shockwave, like an enormous sonic boom, is now tearing away from the impact site through and across the top of the ocean. Larger blocks of debris are still being propelled out of the hole.
7. After a few seconds most of the asteroid and some of the ocean floor have been completely vaporised, and the cavity in the ocean is about to stop getting larger. Most of the world is still unaware that the impact has occurred. The earthquake tremors caused by the impact are moving at a few kilometres a second away from the impact, and at greater distance will produce merely a deep rumbling.

– when the time came – the first complex animals. Life on Earth would have been under threat from cometary and planetismal bombardment, cataclysmic volcanic eruption, solar and cosmic radiation, toxic gases and glacial cold, but not all at the same time, nor everywhere at the same time. Life had a foothold, it had the mechanisms for survival and replication and it had, so to speak, all the time in the world. It had somewhere to go. It could take charge and begin to adjust the world to its liking, and where it could not do so, it began to adapt to fit in with the planet's most insistent demands.

Fast forward more than three billion years and we pause to examine a world so rich in microbial life that – always supposing you could collect the planet's most important tenants and

---

8. The ocean floods back into the cavity in the ocean to cover up the exposed ocean floor. The flood is like the closing of the Red Sea on the army of the pharaoh, and is so enormous that the waters overshoot to form a pillar of water over ten kilometres high, that collapses back producing a giant wave of kilometre scale that expands outwards like a behemoth ripple on a pond. The surface of the ocean at the impact site bobs up and down, producing increasingly smaller waves that follow behind the first.

9. The giant wave, the tsunami, expands outwards across the surface of the ocean at many hundreds of kilometres per hour. With each passing second, however, the height of the wave becomes less.

10. After an hour the wave reaches the coast and rides up the gentle slope towards the shore, becoming taller by the second. At one place on the coast the wave may be a steep roller a kilometre or so in height, at another it is like a rushing tide that sweeps inland. It tears over the land, ripping through forests, cities and towns as if they were made from paper. In places it travels hundreds of kilometres inland over the flat floodplains, over others its advance is held back by mountains.

11. Over the next few hours the wave creeps along the coastlines of over half the surface of the world, penetrating less and less over the countryside until finally, at an enormous distance from where the asteroid collided with the ocean, it becomes no more than an unusually large roller.

12. Over the next few days and weeks the dust and water vapour thrown into the uppermost atmosphere expand outwards and mix with the air currents. Like an enormous volcanic eruption the dust begins to produce unusual weather, and there is a noticeable drop in temperature. Many of the crops in the fields begin to fail, and a planet with only two days' reserves of food begins to face the prospect of global starvation and mass mortality measured in billions.

assemble them amicably in one place – it would form a thick smear of bacterial goo over the entire planetary surface. These microbes are divided into three kingdoms, and one of these kingdoms, the Eukaryota, can be thought of as a tree or shrub with thousands of branches, shoots and twigs, and from one of these twigs are descended all the plants, insects, reptiles, birds and mammals of the modern world.

Nobody knows how many species of complex creature stalk the planet now – estimates range from seven million to seventy million – but these still only represent a small fraction of all the species that have ever lived since complex life first appeared, from the first cooperative ventures of the microbes, more than five hundred million years ago. A few of these creatures, and perhaps more than a few of the microbes, have changed little or hardly at all during five hundred million years of evolution, but the record of fossils in the rocks suggests that any single species can expect to survive only a limited time on this planet: a couple of million years, maybe ten million years, before being written out of the adventure, like characters in a soap opera who have begun to exhaust their potential and exasperate their audience. The story-so-far of life on Earth is tentative in outline, capricious in detail and constantly being rewritten, updated, revised and amended. The authorship is a matter for eternal dispute. Pages have been defaced or erased, whole chapters have been torn out, whole casts of characters appear, begin to take action and then disappear, and although a preface and a chapter one can be inferred, nobody knows how the narrative began.

There is, however, an ending: in fact a choice of endings. Life could go on maintaining the planetary atmosphere at a comfortable temperature for at least a hundred million years, and probably a lot longer than that. But the Sun, forever converting its store of hydrogen into helium and other elements, would continue to grow imperceptibly hotter with time. The silicate rocks would continue to weather, and – with and without life's helping hand – incorporate carbon dioxide into their substance. These rocks would eventually become buried deep in the crust and eventually absorbed into the planet's mantle, and the carbon dioxide would be released again into the atmosphere from volcanic discharges and deep ocean vents. But nothing is forever, and at some point the carbon dioxide levels in the atmosphere would become too low to support the Earth's forests and grass-lands, photosynthesis would come to an end, and all complex life – the creatures that consume foliage, or consume the herbi-vores that depend on vegetation – would begin to fail. The temperature would begin to rise more rapidly. In time – it could be 1,500 million years – global mean temperatures could reach 50°C, and since this is a mean temperature, some parts of the planet would begin to boil. A steamy atmosphere would increase the ambient temperature still further. But water vapour is a vulnerable thing: molecules of water can be severed by ultra-violet light – the process is called photodissociation – and the hydrogen now trapped in water would be free. The hydrogen circulating in the stratosphere would begin to escape. It could take 2.5 billion years to do so, but eventually all of the planet's water would boil into space. Without water, there can be no life of any kind. Volcanoes would continue to release water vapour and carbon dioxide, but without oceans to lower planetary temperatures no rain would ever fall; the hydrogen from water would continue to escape into interplanetary space, but the atmosphere would start to become composed of carbon diox-ide, and planetary temperatures would begin to soar inexorably.

Eventually, the Sun – a main sequence star with a predictable lifespan – would enter its death throes, swell up into a red giant and incinerate the Earth. And that would just about wrap it up for the only home we've ever known. By then, of course, humankind will have long since moved on or died out. Demolition of the now vacant and undesirable residence and the recycling of its fabric will be inexorable. One day, it will all be just so much wafting gas and isolated dust, drifting in galactic space, each atom or molecule the potential raw material for yet another future star, or its orbiting panoply of planets.

NINE

The Solar System

Places in the Sun

As I write, it is raining on Titan. Rain might be too strong a word. It could be a series of light, sporadic showers, or perhaps a fine, soggy mist. But moisture is falling on the western foothills of the continent of Xanadu, and then trickling back down the slopes, eroding drainage channels and collecting in rivers before flowing into lakes and seas, to evaporate once again in the feeble morning sunlight, and rise again with the winds towards the yellow skies of Titan, to form clouds, and to fall once more.

The words that begin with the preface 'As I write' are based not on direct and immediate evidence, but on an assumption. Astronomers have detected on Titan what they call 'widespread and persistent morning drizzle'. They have also recorded 'nearly global' cloud cover at high elevations. On Earth, the cycle of evaporation, condensation, precipitation and erosion shapes the planet's landscape and drives the engine of climate, and does so every day. If you see these things happen once and leave their mark on an already marked landscape, then you could reasonably make the assumption that they go on happening. The difference is that at any time, you can demonstrate that rain is falling, has fallen or is about to fall somewhere on planet Earth, and you can see where the rain has been: you can identify the eroded hills and landslides, the screes, the floodplains, the gullies, ditches, wadis, swallets, creeks, dendritic flows, deltas, swamps, shorelines, high-water marks, mudflats, reefs and sand-banks that provide permanently shifting evidence of falling, flowing and evaporating water, of melting snow and thawing ice. You can, of course, also just go out and stand in the rain. Titan, however, is not such an easy spot for a quick, high-probability weather forecast.

Titan is 750 million miles from Earth, a satellite of the planet Saturn. It is a precise address in the solar system, which is itself an address, a place in the heavens, a round for a cosmic post-man. 'This little canton, I mean this system of our Sun,' wrote Locke in his *Essay on Human Understanding* in 1690, and he certainly included Titan in this, because he later defined the solar system as the Sun, its planets and comets. But Titan was just a vague address until 15 January 2005, when a European lander called Huygens dropped through its opaque atmospheric shroud, and photographed hills, slopes, drainage channels and shorelines as it swung from its little parachute towards the satel-lite's mysterious surface. And the conditions on Titan are noth-ing like the conditions on Earth. The average temperature is minus 180°C. Any water would be forever frozen, as hard as flint, and as nourishing as diamond. The landscape of Titan is covered in a scum of hydrocarbons: ethane, methane and other compounds. At minus 180°C methane has an unusual set of properties. It can exist as liquid, solid and gas. Water, too, can metamorphose into any of these three states, at or around its freezing point, ninety-three million miles or 150 million kilo-metres from the Sun. But Titan is not ninety-three million miles from the solar system's central heating installation. It is more than nine hundred million miles, or 1,350 million kilometres, from the Sun's morning light, its noonday blaze, its twilight's last gleaming. The marvel, the thing that astonishes anybody who can remember the time before the space age, is not that it rains methane instead of water on Titan. The marvel is that – for just a few brief hours – humankind was able to see evidence of that gentle chemical drizzle, and hear the whistle of the winds, and map a vista of distant hills and lakes on a moon known, for 350 years, only as a dull speck around a much brighter speck, visible by telescope only on a clear night.

Saturn has its place in the making of the scientific revolution: Galileo turned his primitive home-made telescope on it four hundred years ago, and was startled to see something he described as 'not a single star, but three together, which almost touch each other'. He was, of course, the first person to observe Saturn's rings. That was in 1610. Two years later, to his surprise, the other two objects, the things that seemed to be 'handles' on the side of Saturn, had disappeared. So he was also the first person to observe what astronomers now call a Saturn ring-plane crossing: he saw the rings edge on, and therefore effectively invisible. Four years later, he looked again, and recorded 'two half ellipses with two little dark triangles in the middle of the figure and contiguous to the middle globe of Saturn, which is seen, as always, perfectly round'. He could see, though he could not know it, that Saturn's rings had, from his point of view, tilted in a way that exposed the gap between the parent planet and the dramatic disc of icy rubble that orbits it. In 1655, the Dutchman Christiaan Huygens, using a more powerful telescope, correctly decided that Saturn was encircled by a ring that never touched it. In the same year, he spotted Saturn's moon Titan.

Saturn – even in Galileo's day the dimmest of planets or 'wanderers' – continued as a kind of star in the imagination's firmament. Voltaire's *Micromegas* (1752) is a giant from the star Sirius, who meets and befriends a dwarf from Saturn, a trifling thousand fathoms in height: they jump on Saturn's ring, make their way from moon to moon, leap (with all their servants and instruments) on to a passing comet, snatch a lift across the 150 million intervening leagues to the satellites of Jupiter, coast by the planet Mars and its two moons (Mars indeed has two moons, called Phobos and Deimos, but these were not discovered until 1877) and proceed to land on planet Earth, which they walk around in thirty-six hours. They are about to declare it lifeless when (with the aid of a microscope) the Saturnian catches a whale in the Baltic, and then a ship.

Micromegas is a political satire, but it is also early science fiction. It demonstrates that, two centuries before Captain Kirk, Mr Spock and the starship *Enterprise*, Voltaire could assume a readership that was alive to the idea of space travel, of traffic between the planetary systems, and a plurality of worlds, a firmament fizzing with community and identity and variety.

There were places to go, and people to see, and things to learn at each planetary address. Immanuel Kant in 1755 proposed that intellectual stature might advance according to distance from the Sun, which would make Mercurians and Venusians the dullards of the solar system, place Earthlings on the middle rung of the ladder of beings, and endow Saturnians and Jovians with the brilliant intellects of superbeings, all products of Nature's unlimited fruitfulness. Humphry Davy, in his last, and posthumous, book, called *Consolations in Travel* (1830), imagined Saturnians floating in the gas giant's dense atmosphere. In 1837, in a book called *Celestial Scenery*, Thomas Dick, a Scottish clergyman, tried to calculate the potential population of the solar system. Assuming the average density in England to be 248 souls to the square mile, he proposed that 5,488,000,000 beings could find homes and a living on the surface of Saturn. He also found room on its rings for another eight trillion inhabitants.

Some of this fervent debate about life beyond Earth began in a religious century and had its roots in wider religious argument. 'If relations were opened up between ourselves and these planetary men – whom M Huygens says are not much different from men here – the problem would warrant calling an Ecumenical Council to determine whether we should undertake the propagation of the faith beyond our globe,' wrote the philosopher Gottfried Leibniz in 1704, in an essay not published until 1765, long after his death. Some of this debate also had its roots in a different kind of reasoning. 'If all places to which we have access are filled with living creatures, why should all those immense spaces of the heavens above the clouds be incapable of inhabitants?' Isaac Newton asked, and three centuries on, the question has not changed, although an answer seems as far away as ever.

The point of this detour through European science history is a simple one: at the dawn of the space age, almost all of the eighteenth-century questions about the other citizens of the solar system remained open. Telescopes improved, physics became more sophisticated, scholarship became more systematic. But in October 1957, when Sputnik 1 was launched by the Soviet Union, purely to make a triumphant but monotonous telephone call back to Earth, and along with it a geopolitical point to the peoples of the western hemisphere, astronomers and physicists were ultimately no more certain about life on other planets than Newton or Voltaire. That is, despite three hundred years of careful study, they conceded that life might exist on the other planets: no one could be sure.

C.S. Lewis in 1943 in his book *Perelandra* confidently made Venus a warm, watery world thick with primordial foliage and paradisiacal innocence. He was composing a science fiction allegory, but his Venus was as plausible as any other writer's. Eight years later, the giant of space science and space fiction, Arthur C. Clarke, proposed in *The Exploration of Space* that there was 'no foundation at all for the common idea that Venus is a world of oceans and steaming swamps, like our own planet some thousand million years ago', but he did not rule out the possibility of intelligent life, and he allowed himself to speculate about what kind of cosmology a Venusian might be able to construct, beneath the dense envelope of gas that veiled the second rock from the Sun. Clarke did however expect plant life of some kind on Mars, 'though there is no evidence at all for animal life or intelligence'.

The first Apollo astronauts to return from the Moon in 1969 were detained in quarantine for sixteen days before they could embrace their families, because of fears that they might have picked up a sinister lunar infection, and might introduce it accidentally to the rest of the human race. When the Viking robot mission to Mars set off in 1976, Nasa scientists released an artist's impression of what they imagined might possibly be life on Mars: fungal creatures whose tissues were based on silicon rather than carbon, immensely slow-growing, probably slow-witted, but at least alive: creatures that could multiply from a single cell, consume resources, metamorphose the elements, replicate themselves and ultimately die and decompose.

It was the Viking mission that appeared, for the first time, to settle the question. By 1976, little automated emissaries from Earth had visited Venus, and observed at close quarters a planet cloaked in clouds of sulphuric acid, with atmospheres as oppressive as the ocean floor, and ground temperatures hot enough to melt lead. It was no place for life. When in 1976 a Viking lander dropped down on the surface of Mars, it appeared to settle that question, too: the Red Planet was a dead planet, waterless, inert and masked by a veil of gas so tenuous that it couldn't reach the planet's own mountaintops.

That was it. There were no neighbours. Life in the solar system, at least, was confined to a ball of rock just big enough to keep itself warm and active and retain an atmosphere that extended for perhaps a hundred kilometres. Beyond that, there was no life at all; at least until November 1957.

The first Earthbound creature to make the great leap into the unknown was a dog. Laika was a stray, a mongrel rounded up on the streets of Moscow: she had no name, and the scientists who trained her for stardom called her a variety of distant or half-affectionate things before settling on Laika, which means Barker. They strapped her into Sputnik 2 and launched her from Baikonur Cosmodrome on 3 November, in time to mark the fortieth anniversary of the Bolshevik revolution and the bloody birth of the Soviet Union. Laika was a dog selected to make a political point: Soviet scientists had stunned the world – and shaken the US – with Sputnik 1, a basketball-sized lump of metal that did nothing but emit a steady beep as it girdled the globe every ninety minutes or so. But Laika was a message, an announcement that the Soviet Union could do more: it could pack scientific instruments, a life-support mechanism, food, a sanitary system and a passenger into one small spacecraft and send it far above the filmy envelope of the atmosphere, into the blackness of space under the brightness of the stars. In 2002, one of the scientists involved in Sputnik 2 revealed the mournful truth at an international conference: the life-support system had not worked for long, and Laika had died within hours, probably from overheating. Laika may have circled the world at eight kilometres a second four or five times before she closed her eyes forever, but her corpse went around the world more than 2,500 times before the spacecraft and its experimental passenger perished in a little fireball on re-entry into the atmosphere in April 1958.

Laika may have died in vain – the experiment did not reveal enough useful information, and animal launches continued – but she became probably the most famous dog in the world, or more properly out of it: the dog star, the first hound of heaven. She has remained far more famous than Strelka and Belka. These two – Little Arrow and Squirrel – went into orbit in

August 1960, on Sputnik 5. They were not alone. They flew into space with forty mice, two rats and some plants. After a day in orbit, they came down again and landed safely. So Strelka and Belka really did make history: they went up, came down, and went back to a dog's life. They may even have galvanised the United States into joining the space race. 'The first space satellite was called Sputnik, not Vanguard. The first country to place its national emblem on the Moon was not the United States but the Soviet Union. The first passengers to return safely from a trip to outer space were named Strelka and Belka, not Rover or Fido,' said Senator John F. Kennedy, presidential candidate, at an address in Dallas in 1960. He went on: 'The only thing that will deter Mr Khrushchev from loosing his hounds on us will be a strong America.' The Soviet president, Nikita S. Khrushchev, loosed at least one hound on America anyway. He presented one of Strelka's six puppies to Caroline Kennedy, the little daughter of the new president.

The space race had taken off. Humankind had begun to prepare for the kind of voyages foreseen by Voltaire, calculated leaps from one planet and planetary system to another, journeys that for humans would begin in April 1961, with Yuri Gagarin's one, momentous orbit of the Earth. The most distant journey, at the time of writing, and for the foreseeable future, has been to the Moon, 240,000 miles away: between 1969 and 1972, twelve humans stepped onto the Moon, collected 382 kilograms of its rocks and dust, planted a few instruments on its surface, abandoned some technological litter, and then sped away. The last visitors, the crew of Apollo 17, left to the music of the Carpenters, singing 'We've Only Just Begun'. In fact, that was the last human flight beyond low Earth orbit. But the exploration of the neighbourhood would continue with a series of ever more remarkable automatons that would, as proxy humans, begin to explore all the way to a region known as the heliopause, an ill-

defined, shifting frontier that represents the limits of solar jurisdiction. These last outposts of the empire of the Sun are where
the pressure of the solar wind is matched by the pressure of
radiation from the surrounding stars: the place where interplanetary space begins. The heliopause encloses a very tiny spherical
volume of real estate when compared to the galaxy of which the
Sun is itself but a tiny light, but for the citizens of the third rock
from the Sun this space within is still unimaginably large, and
difficult to comprehend.

The simple way to make sense of the solar system is with a
little diagram. Massive Sun, small Mercury, bigger Venus, very
similar Earth, smaller Mars, a space filled with broken-up bits of
planet called the asteroid belt, then a huge gas giant Jupiter, then
another called Saturn, then Uranus, then Neptune and then tiny
Pluto, except that sometimes Pluto is nearer than Neptune
because its orbit is more elliptical. The convention is to draw
them roughly to scale, with all the planets strung out in a line to
one side of the Sun. This however, is simply a way of remember-
ing the planetary order in the journey from the centre of the
solar system to its most distant rim. But the nine planets – for
astronomical pedants, the correct description would now be
eight planets and the nearest minor planet, because Pluto is not
the last planetary body, nor even much of a planet – are all in
orbit around the Sun, and travelling at different speeds along
those orbits, so grand alignments of the kind routinely demonstrated in a simple drawing hardly ever happen.

The other conventional way to think about orbiting planets
is to calculate, build or imagine an orrery. This is a coined word
for a clockwork model of the solar system, originally built for an
aristocrat, the Earl of Orrery, in 1713. It is a geared instrument,
again with the Sun in the centre, and with little brass or painted
representations of the planets revolving around it, showing
where Mars or Jupiter or Saturn might be at any time in relation

to the Earth. An orrery, however, is never to scale. If it was, then either you would not be able to see the whole working system at a glance, or some of its components would be so small you couldn't see them at all: they would be smaller than the resolving power of the human eye. If you could scale the solar system down by a factor of a million, so that planet Earth was roughly the size of a small cottage in Hastings, southern England, and the Sun was the size of the Cities of London and Westminster, then Saturn would be an object the size of a sports stadium near Zagreb in Croatia, and Pluto would a bus shelter in Pakistan.

Even this imagery won't help. Zagreb stays where it is, in relation to Hastings, England. But Saturn is moving in orbit around the Sun. So, of course, is the Earth, but being nearer the Sun it moves at a much greater speed. At this point, it doesn't make much sense to use terrestrial imagery, because it would require you to imagine someone in a house in Hastings, orbiting quite swiftly around central London, trying to keep track of a sports stadium first observed in Croatia, but moving much more slowly along a track that would take it through Hungary and Poland, out across the Atlantic and eventually back to Croatia via the Sahara. In these circumstances, models are no help: we try to reduce the machinery of the heavens to terrestrial scales, but this just makes it more difficult to imagine. The solar system is so enormous as to justify the term 'unimaginable': the marvel is that people *have* imagined it, and imagined it very accurately, accurately enough to plan journeys to the most distant planets, and most of these journeys were imagined long before the days of high-speed computing, and some of them were imagined even before the revolution in physics launched by Albert Einstein. The heirs of Galileo and Kepler used a few mathematical principles to confidently predict the day, the hour and the minutes of the transit of Venus across face of the Sun, an event calculated to happen only eighty-two times in six thousand years, and not observed at all until 1639, when it was seen by just two people, one of whom knew it was going to happen, and was ready to observe. They acted upon the principle that the machinery of the heavens performed according to a kind of celestial clockwork, and the point about clocks is that you can set your watch by them, in every sense of the term.

The transit of Venus sounds like one of those bits of astronomical terminology that could be safely left to the experts, along with azimuth and perihelion and the precession of the equinoxes. But the transit of Venus changed all our lives. Because it could be observed from different parts of the world by different teams of observers, the passage of a neighbouring planet moving directly between the Earth and the face of the Sun could give astronomers a measure of the precise distance between the Earth and the Sun. They had already begun to calculate the relative distances between the planets, using the astronomical unit, or AU, the interval between the Sun and the Earth, as a yardstick. But the transit of Venus allowed them to mark precise notches on the yardstick. It was to observe a transit of Venus that Captain James Cook carried Sir Joseph Banks and an astronomer to Tahiti in 1769. Cook's Admiralty orders commanded him, after he had observed the transit from the Pacific, to discover the southern continent that for hundreds of years Europeans had conjectured must exist south of the Equator. Cook quickly found that the Pacific was mostly ocean, but he explored it anyway. His three voyages were not just models of navigational thoroughness that set new standards for all succeeding exploration: they opened a new chapter in world history, and began to set a pattern for the colonisation of the South Pacific.

Cook discovered neither Australia nor New Zealand, but he made the maps and named the landmarks and asserted the primacy of his nation. So I grew up in a place given the name of Devonport, Auckland, knowing the date of the battle of

Hastings, because somebody in Lancashire in the seventeenth century predicted the transit of Venus, and somebody in London in the eighteenth century thought it was important to observe this transit from the far side of the world. The city of Auckland was named after George Eden, Earl of Auckland, a First Lord of the Admiralty and a patron of the city's founder, Captain Hobson. Devonport in Auckland is the site of the New Zealand naval base, and is named after the British naval base at Plymouth. Our addresses – the locations, and the names that we give them – are chance events, but the chain of chance is a long one, and one link in this chain was the drive to understand the order of the planets in their silent concourse around the Sun.

This drive goes on. Even before Yuri Gagarin of the Soviet Union made his one, famous, ninety-minute orbit of the planet in April 1961, Soviet and American engineers had tried twenty-one times to send robot missions to the Moon, Mars and Venus. By the time Neil Armstrong and Buzz Aldrin stepped onto the Moon's dusty surface in July 1969, the two competing superpowers had launched or tried to launch another ninety missions to other heavenly bodies. Since then, and up to the time of writing, the world's space agencies have dispatched another eighty spacecraft to orbit the Sun, explore the other planets and their moons, or sail past or smash into comets, to meet asteroids or to catch fragments of superheated, high-speed shrapnel hurled across the solar system by explosions on the surface of the Sun, and this count does not include the space-based telescopes that measure the radiation of the heavens, nor the manned flights aboard Skylab, or Mir, or the space shuttle.

Each of these missions had a solemn, overt commitment to quantifiable scientific discovery, or to the resolution of a precisely framed scientific question, or more usually both. But in every case, the sponsors, and the begetters, and the researchers who would later comb through epic quantities of data from these missions understood that behind each expensive, highly defined journey there hovered a set of bigger, more imprecise questions: was there life out there, once upon a time if not now? Could there ever have been life? Could life beyond Earth one day be possible? If there were no neighbours in the deserted villages of the solar system, could humans at least convert one of the empty dwellings into a second home? If conditions out there were too terrible for life as we know it, then might we begin to look for life as we did not know it? And if the answer to all these questions is no, then how could life have happened on Earth and nowhere else? What unique, wonderful, miraculous chain of circumstances took the raw materials common to the whole of the solar system and confected from it light-seeking, feeding, growing, replicating organisms that eventually evolved into sea slugs and Shakespeare, violets and Voltaire, termites and Tony Blair? In a sense, every mission to another object in the solar system was really a question about the Earth. There is only the smallest shift of emphasis from the very precise question 'Why are we *here*?' to the larger question 'Why are *we* here?' The difference is, of course, that perhaps we can answer the first question. That does not stop us from wondering about the second.

Right now, there are two tiny spacecraft – each the size of a Volkswagen Beetle – exploring the farthest frontiers of the solar system. Voyager 1 and 2 were launched in 1977, each with three onboard computers equipped with memory that, at the time, must have seemed powerful: in fact the total word capacity of all six computers is about 32,000, and the total memory about 68Kb. That computing power now seems derisory – the laptop on which I am writing these words has more than a million times the capacity. These instruments had to navigate the spacecraft, maintain its attitude, detect and correct internal faults, control its internal economy, operate its camera, manage its thrusters, monitor its scientific instruments and keep in regular contact with the engineers back on Earth, now billions of miles away. To store and relay scientific data, each spacecraft also carried tape recorders, and by the time Voyager passed Neptune these tape recorders had played back to Earth enough data to fill six thousand sets of the thirty-two-volume *Encyclopaedia Britannica*. Each spacecraft was directed first to Jupiter and its moons, and then to Saturn and its satellites, and then Voyager 2 went on to Uranus and Neptune.

To save fuel, each stole some extra thrust from the planets it sped past: this is a manoeuvre called gravity assist, or gravity slingshot. The logic of Newtonian physics says it can be done, as long as the total energy of the solar system is conserved. So when each spacecraft sped by Jupiter, it gathered speed by slowing the planet down. Because Jupiter is enormous – 318 times more massive than Earth – it casually flicked the spacecraft onwards as an elephant might flick a wisp of straw. To be precise: each spacecraft initially doubled its speed by passing Jupiter, and Jupiter slowed in its orbit around the Sun by about one foot per thousand billion years. Voyager 2 also borrowed speed from Saturn and Uranus, and in doing so cut its journey time to Neptune by twenty years. The journey is not over. Each spacecraft is now pushing at sixteen kilometres a second or more towards the farthest limits of the Sun's domain, past the point where the fine spray of particles from the Sun is balanced by the distant sniping of fragments from the nearest stars, and through the cloud of comets believed to patrol the perimeter of the solar system. Around 40,000 years from now, each will reach the gravitational domain of another star.

The adventure, conceived before the first landings on the Moon, and authorised in 1972, was designed and planned with 1970s technology to answer 1960s questions: what happens on the moons of Jupiter, and what are the rings of Saturn? The engineering was meticulous, and the scientific payload assembled as economically as possible, but each spacecraft not only beams increasingly faint messages back to Earth, it bears a to-whom-it-may-concern letter of introduction that might just possibly one day reach some distant, faraway civilisation tens of thousands of years in the future. Both spacecraft carry what became called 'golden records' – gold-plated copper discs rich in the sounds, music and languages of planet Earth: Chuck Berry and Beethoven, whale calls and birdsong and the sound of surf, and greetings in fifty-five Earth languages. With each grooved sound-portrait of the home planet is a cartridge, a needle and symbolic instructions on how to play the record, no doubt in the expectation that a sufficiently advanced alien civilisation would by that time have itself discovered long-playing records and high-fidelity equipment. Never mind the comedy inherent in dispatching, for the astonishment of an extraterrestrial, a technology that many young Earthlings would no longer recognise, what matters is that the US space agency Nasa, the scientists behind the enterprise, and the society that bankrolled them, were still prepared to believe in life beyond the Earth: that was the point of a record saying 'Hello!' in fifty-five languages.

But even the hard-nosed scientific work done by the Voyager spacecraft on their way to the stars began to raise questions about life beyond the last wisps of the Earth's atmosphere. Voyager sped past Jupiter's moon Europa, and photographed a smooth coating of surface ice remarkably free of the fractures and impact craters that pockmark other moons and planets in the solar system. The easiest explanation for this is that Europa has beneath its ice a vast, deep ocean through which water upwells and freezes again each time an asteroid or a lump of comet hits the surface. If Europa has a liquid ocean below its surface, then it must also have a source of heat at its core: how else could water stay liquid so far from the Sun? If a planet-sized body has a source of heat, and liquid water, then it has the two essentials for life. Nobody knows quite what to do next: how do you design a mission to cross more than six hundred million kilometres between Earth and Jupiter at their nearest, land on an icy surface of a Jovian moon, drill through ten kilometres or more of rock-hard ice, and lower a robot submarine, filled with fuel, instruments, searchlights and television cameras, and hundreds of kilometres of wiring, into the unknown? Voyager also took the first-ever close look at Saturn and its rings: astronomers had guessed at their fabric and the forces that maintained them, but it is one thing to calculate, quite another thing to see photographs taken at relatively close range, of thousands of strands of ice particles and small icebergs, arranged in a disc 200,000 kilometres across but less than one kilometre thick. Voyager sped by Titan, and confirmed that it was indeed not unlike a rocky planet, with a dense atmosphere of nitrogen along with methane, ethane and other gases. People said at the time that it could have been a model for the early Earth; perhaps it might one day become an Earth-like object, in the last failing days of the solar system.

Billions of years from now, when the Sun swells to a red giant, and burns Venus and Earth to cinders, where could survivors flee to? Titan, rich in hydrocarbons, might be a very different place once its frozen water begins to thaw. Mars might be a dead planet, but was it always dead? Its atmosphere is so tenuous that its highest mountain, Olympus Mons, twenty-four kilometres high, actually sticks out into space. Its pressure is so low that no liquid water could survive on its surface: it would flash to vapour and leak into space. Its average temperature is minus 55°C, and it is masked for months by furious dust-storms. But even though Mars is now more arid than Earth's harshest desert, its surface bears tantalising signs that things might once have been different: depressions that look like lakebeds, slopes that look like beaches of inland seas, dendritic patterns that look exactly like dried-up riverbeds. So the search, on Mars, is for the evidence of life, however primitive that might once have been.

The search on Titan is for something else: the conditions from which life might emerge. Thus the mission to Titan was also a mission back to what might have been the past of the primitive Earth. To make the journey, a planetary probe the size of a washing machine called Huygens enjoyed a seven-year piggyback aboard a much bigger spacecraft called Cassini, which at the time of writing is still touring the moons of Saturn and sending back stunning imagery.

On 15 January 2005, Huygens began and ended its career with a lonely, once-only plunge into the dense haze that always masks Titan: it survived long enough to photograph the land-scape, record its winds, measure its temperatures, detect its gases and report, for the hundredth of a second of its remaining life, on the texture of the terrain it touched, all the while dispatching data back to the mother ship Cassini, before it went silent forever. Teams of scientists and engineers from all over Europe and America had schemed and dreamed and designed and re-designed and checked and checked again, some of them invest-ing more than a decade of a research career on this one brief throw of the celestial dice; some of them with no guarantee that they would still be scientists when the answers came in, if they ever came at all.

Journeys such as Cassini's are almost acts of faith. The space-ships cost hundreds of millions of dollars to design, build and launch; the journeys take years, and they are conducted in an environment so cold, harsh and hostile that the spacecraft's survival is itself a source of wonder. The questions that Cassini-Huygens was designed to answer were formulated in the 1980s; the spacecraft was designed and fitted with instruments from the 1990s; and the first and only reports from Huygens arrived in one short burst on one night in 2005. The pictures that blew in, almost with the wind, along with all the other data were incomplete, and far from perfect, and they recorded just one brief moment at one unrepresentative address on a satellite of a distant giant planet. But they were also somehow miraculous: they revealed an utterly alien world, made of unearthly materials, a world that somehow nevertheless looked like Earth. The cameras recorded no rain, but they did record clouds and hills and river-channels and what looked like a shoreline and – as the camera hit the ground – eroded and worn boulders of frozen stuff that were shaped like cobbles from a riverbed. Cassini went on beyond Titan, but came back again and again, and gradually, through the murky atmosphere and the bands of clouds that form at mid-latitudes, and with at last some idea of what might be going on underneath them, astronomers began to see shapes and patterns: among them a Titanic continent or plateau the size of Australia that seemed to be of purer ice. If this plateau was weathered, pitted, scarred or sculpted by falling moisture, then perhaps it also contained caves. With the poem 'Kubla Khan' by Samuel Taylor Coleridge in mind, they named it Xanadu. They also detected a darker area nearby, perhaps a plain, that for want of any better connection, they dubbed Shangri-la.

And on the day I began this chapter, astronomers based in California, using near infra-red images from telescopes on Hawaii and high in the mountains of Chile, saw, or thought they saw, rain that had dissipated before 10.30 a.m. local time. They described the kind of faltering precipitation that happens in Hastings and on the high Downs of Sussex, near the bush-clad hills to the north-west of Auckland, on the western slopes of the Rockies and the Andes, from moisture-laden clouds that had been pushed up the slopes by the winds, to condense into coastal rain. Titan is not like Earth. Its year – one orbit around the Sun

– is equal to almost thirty Earth years. It takes Titan sixteen Earth days to rotate upon its axis, so 10.30 a.m. on Titan would be about three Earth days after sunrise, and the cold, feeble light from the faraway Sun would cast only bleak daylight and register only a flicker of a rise on any Titanic thermometer. And at an average of minus 183°C, what fell could never be water.

Titan may be distant, but if it experiences 'widespread and persistent morning drizzle', then suddenly it becomes less far away. This long-range weather forecast establishes an unexpected intimacy. Titan is uninhabitable, but if it were inhabited, Xanadu's smallholders and crofters would have at least one thing in common with Earthlings from southern England or north Wales or western Scotland. They could talk about the weather.

TEN

The Galaxy

There Goes the Neighbourhood

Take a good look at the night sky. It won't be there forever. A cosmic slum-clearance programme is on the way, or maybe a galactic city-centre renewal initiative, or perhaps a redrawing of the celestial boundaries. Right now, and travelling at colossal speed, the Andromeda galaxy is heading for a collision with the Milky Way. Quite when this will happen is uncertain – in two billion years, or maybe three, or even more – but it could certainly happen within the lifetime of our parent star, the Sun. The future shape of the Milky Way is uncertain, but if a tight cluster of several hundred billion stars, spinning around its own galactic centre, grazes against, or perhaps crashes at 140 kilometres a second into, another tight cluster of maybe a thousand billion stars, also pirouetting around its own centre, there is bound to be some damage, or at the very least a substantial rearrangement of the existing galactic fabric.

There is a measurable chance that the Sun and all its planets, already in the outer suburbs of the galactic metropolis, will be dislodged, and sent spinning to the furthest fringes of the celestial conurbation. There is even a very small chance that the Sun will be detached altogether, and become part of Andromeda as it brushes past. Anything is possible. The prediction is confident, but at the time of writing only recently made. With a time scale measured in billions of years, the details can wait. It seems,

179

however, that an encounter is inevitable. Either this property is condemned, or the neighbourhood is heading for substantial alteration and enlargement, accompanied by some demolition, or some parishes will find themselves transferred to another administrative district.

Andromeda clouds our future, just as it has a nebulous place in our past. It has already told us something about our immediate neighbourhood. It has always been one of the fixed stars. What we see, after dark, when the clouds part, and far beyond the Moon and the planets, are the fixed stars. They are part of our geography. They are not, of course, fixed: they move in relation to each other, and we move in relation to them, but the distances to even the nearest of them are so colossal that we detect no change in their relative positions, and so, to us, Orion and Cassiopeia, the Pleiades and Sirius the Dog Star, Andromeda and the Southern Cross, seem to parade in drilled and inexorable order across our night sky, an order so meticulous that the precise pattern overhead is enough to tell us the hour of night, the season of the year and even our place on the planet.

These heavenly features are only waymarks, signposts to the real action at the heart of the galaxy, and on a clear, moonless night we can see the celestial equivalent of downtown: a faint, wide ribbon of whiteness that reminded the ancients enough of spilt milk for them to name it in Greek *Galaxias*, and in Latin *Via Lactea*, the Milky Way. For all practical purposes, this was our cosmos. We now know what we mean by galaxy: a world of stars, a gathering, a coming together, a mutual gravitational embrace of stars, stardust and stray atoms, supposedly swirling around an unimaginably massive black hole at the centre, glued together by something unknown. The black hole is a conjecture: lots of galaxies have them, maybe all of them do. Who knows? Most galaxies have never been detected, never been photographed.

The unknown glue that extends from the galactic centre to the furthest fringes, the mysterious mortar that gives a galaxy its distinctive architecture, is usually called cold dark matter because no instrument can measure its temperature, and no detector can feel it. But it is there: it must be, because the mass of all the visible stars is not enough to explain the graceful, swirling clusters of billions, even hundreds of billions, of stars that bear the generic label of galaxies. So some unknown and very-difficult-to-know substance makes up most of the galactic mass, and the billions of massive, gleaming things called stars are only a small fraction of this mass. We know this now. My grandfather's generation did not. For him, and for the astronomers of his day, the night sky was all there was: a location, a home address, a skyscape of light extended indefinitely, but it remained one vast mass of stars, somehow contained in a thing called a universe. The first shamans, seers, priests, navigators and astrologers probably began making mental maps of the night sky in the Old Stone Age. Their inheritors almost certainly erected the pyramids of Yucatan and the temple of Karnak and the megaliths of Stonehenge as astronomical devices that marked the cycles of heaven. Such devices served to measure the solstice and equinox, the passage of the seasons, the journey of the Sun through the zodiac and even the direction a pharaoh's soul might take after his death. Ptolemy of Alexandria, 1,900 years ago, decided that the Earth was the motionless centre of the cosmos, and about it rolled a series of spheres that enclosed the Moon, Venus, Mars, the Sun and the other visible planets, and beyond them was the sphere of the fixed stars; this provided the topography seven hundred years ago for one of European culture's great voyages of discovery, Dante's *Divine Comedy*.

Galileo, four hundred years ago, knew that the Earth did move. He turned a telescope on the Moon and the planets, and began to see the heavens differently, but he too saw the stars for what they were: fixed. They wheeled overhead, the pattern of their movements might change imperceptibly in the course of thousands of years, but their relationship to each other seemed to remain fixed. Telescopes improved, and for the first time astronomers began to see new stars beyond the familiar ones, and sometimes quite unpredictably a new star – a nova or supernova – might appear. As telescopes became more sophisticated, and their mirrors more perfectly polished, astronomers saw further and more clearly, and the tally of stars, which even three thousand years ago seemed numberless, continued to soar. But the big picture remained the same: the stars looked as though they were fixed. Of course they moved – the galaxy had a centre, and the Sun and other stars moved in relation to it – but they moved at the same speed, and maintained the same order, and somewhere overhead was evidence that the galaxy had a heart, and that the faint splash of spilt milk high in the heavens at midnight was in fact a density of stars so huge as to seem impenetrable, in the direction of the heart of the galaxy.

Everything that astronomers could see remained within a secure order, and right up until the first decades of the twentieth century the night sky provided a kind of surface picture of cosmic reality: what you could see with the naked eye was only the immediate detail, but it was nevertheless a simple representation of everything that could be seen. Ever more remarkable instruments could see deeper and further, revealing numberless stars, extending for tens of thousands of light years. Nevertheless, the big picture was the one that Ptolemy and the pharaohs and Galileo had seen. The fixed stars were the fabric of the galaxy. Stars might be born, stars might die, but the big picture did not change. The galaxy was all there was. The galaxy *was* our universe.

Then, in 1923, Edwin Hubble, working at Mount Wilson in California, turned his hundred-inch telescope on the celestial smudge known for more than a thousand years as Andromeda, and by astronomers as M31, Messier object number 31. By this time, Hubble and everybody else knew that Andromeda wasn't a star. They thought it was a nebula, a cloud of glowing gas that became the birthplace of stars. One of the stars in the Andromeda nebula he identified was of a class that astronomers had learned to use as a measure of distance. So he tried to measure the distance to Andromeda. To his surprise, he found that Andromeda seemed to be at least 800,000 light years away. But the Milky Way galaxy itself was only about 100,000 light years across. If his calculations were accurate, Andromeda could not be a nebula. It must instead be another galaxy. That is, beyond this island universe, itself an assembly of countless stars, there was another island universe. In one sense the phrase 'to his surprise' may not be true: perhaps Edwin Hubble already suspected that that there might be more to the universe than the Milky Way, and perhaps discoveries come most efficiently to those who know what they might be looking for.

Nevertheless, 'to his surprise' still seems an understatement. The history of exploration is a history of astonishment, of the discovery of the once-unimaginable, of the sudden transition from the explorer's small world to a much wider horizon. The traveller climbs a peak and suddenly before him is the Pacific Ocean, the biggest single feature on the planet. Of course he is 'silent, upon a peak in Darien' and he stares, like the men with Keats's stout Cortez, 'with a wild surmise'. Keats was mistaken: the first European to see the Pacific was Balboa. But there is no mistaking the extraordinary nature of Hubble's discovery. This was in one way like all discoveries, but in another it had no precedent. He had looked at all there was, and discovered that it was only a tiny, negligible fraction of all that there might be. In a few hours, he had rolled back a screen and exposed a universe infinitely vast and unexpected, and of course he did so not from a peak in Darien, but a peak in California.

Hubble, in the end, identified twenty-three distant galaxies. He also surmised, before he died, that these galaxies were rushing away from each other, as if being hurled apart in a great explosion. So he had planted the idea that the universe seemed as if it had a beginning, and might be moving towards an end. The stars were not fixed. Nor were they, however big, and however colossal the distances between them, of any great consequence, not compared to the immensity of distance between the galaxies. But even though these distances were enormous – far greater than Hubble had first proposed – he had also discovered another important thing about twelve of them. As far away as they were, they appeared to belong together. So Hubble had made a third profound discovery: that you could have groups of galaxies, galaxies of galaxies that were in some way different from other groups, or galaxies of galaxies. He had discovered a new order. He had discovered a higher geography.

Suddenly, to Hubble, and all the astronomers after him, the Sun was a star among millions, but these millions of stars constituted an address: an identifiable position within a conurbation of galaxies, a county or a duchy or a kingdom of galaxies; federated states of galaxies beyond which there were other galaxies and groups of galaxies that were not with us, that were out there, not in here; that were different, foreign and other, intangible, unreachable and unknowable, except in the sense that streams of photons, little particles of light, had finally made it from some faraway and ever-receding source, to bounce from a reflector telescope's mirror and onto a photographic plate at the Mount Wilson observatory.

These ghostly impressions of faraway entities provided not so much a geography of the universe, as a history: since each particle of light had begun its journey ten million, or a hundred million, or a billion years before, it told a story of what had been, and where it had been, rather than what might still exist, and where it might be now.

Even so, Hubble's discovery had added new precision to the address book. The Earth is the third planet from the Sun, part of a solar system which has a location in an identifiable neighbourhood in the Sagittarius arm of the Milky Way galaxy, which is one of a cluster of galaxies known as the Local Group; and this cluster has a place, too, in relation to other clusters: it is on the edge of an identifiable federation called the Local Supercluster, or sometimes the Virgo Supercluster. It takes this name because, at the heart of this federation, more than a hundred million light years in diameter, and home to a million billion stars like the Sun, is the Virgo cluster. There are around two thousand galaxies in the Virgo cluster. The Local Group contains about thirty galaxies. It is one of about fifty groups or clusters of galaxies in the Virgo Supercluster.

It is tempting to search for planet-based parallels, down-to-earth analogies that reduce these awesome scales to a measure that makes sense. But in the scale of the observable universe, home to between one and two hundred billion galaxies at the most recent estimate, the Virgo Supercluster cannot be compared to the European Community, or to the United States, or even to a geopolitical power bloc such as Nato, or the Warsaw Pact countries. In the scale of the universe, the Virgo Supercluster hardly counts as a conurbation, with Virgo as a metropolis and the Local Group as an overgrown village gradually absorbed by its richer, bigger neighbour: greater Manchester, perhaps, with the Local Group as Chorlton-cum-Hardy, and the Milky Way galaxy as a small sweetshop in a neglected corner of Chorlton. Just as Manchester is one city among many in the north of England, so the Virgo Supercluster is just a name on a cosmic provincial map, with neighbours near enough to have proper names: Coma, and Hydra, and Centaurus, and the Perseus-Pisces Supercluster, itself a great wall of clusters that stretches for three hundred million light years, and marks, for us here in the Virgo Supercluster, a kind of celestial dead stop, a cosmic cliff, an escarpment, beyond which there is nothing: a void, an enormous space, a place of darkness, a huge hole a hundred million light years across, called the Taurus Void. And on the other side of this void, there are more galaxies, and clusters of galaxies, and superclusters, and the pattern begins again.

How did we get to this point? How did a life form, grovelling on the surface of a tepid planet in a thrown-together planetary system around an insignificant star in a very ordinary spiral galaxy within a nondescript local group, manage to pinpoint its place in the pan-galactic neighbourhood? The detailed answer involves a dangerous and potentially overwhelming avalanche of astronomer's jargon: Cepheid variables and charge-coupled devices, standard candles and spectroscopes, redshifts and relativity, gravitational lensing and gamma-ray bursters, parallax, parsecs, pulsars and period-luminosity relations, interferometry and infra-red radiation, adaptive optics and Olber's paradox, inertial frames and inverse square law, entropy and elliptical orbits, Halley's Comet, Hubble's Constant, Hertzsprung-Russell diagrams and the hot big bang. The simplest answer of all involves simple words: curiosity, a community of observers and a spirit of cooperation.

Individual astronomers may be cantankerous, competitive and cruel. As a collective, however, they have probably done more for the human race, at lower cost and with less collateral damage, than practitioners of any other single discipline. Notoriously, surgeons and physicians, with their initial doctrines of humours, miasmas and bloodletting, may have killed more people than they cured for the first twenty centuries of medical science. The founders of the world's great religions preached obedience, submission, love, kindness, sobriety and order, but their disciples also killed, burned and tortured each other unremittingly for the next two millennia. For the past four hundred years, astronomers have simply looked up, made notes, shared their data and debated their ideas.

The first scientific astronomers – Copernicus, Galileo and their followers – four hundred years ago, experimented with glass lenses and tubes, devised explanations and printed both their observations and their theories. They made the assumption that what they could see through a telescope was probably a closer representation of reality than anything they had hitherto read in Aristotle or St Augustine. They adopted what is now called the Copernican principle, which says that if there is nothing special about planet Earth, then the physical laws that apply on Earth very probably also apply on Uranus and Aldebaran, Andromeda and Ursa Major. They tried to work out how far away the planets and stars might be, and what they might be made of, and to understand the seeming clockwork mechanism of heavenly orbits. This community of observers, thinkers and debaters was pan-European, and soon pan-global. It began with a handful of educated, privileged and resourceful individuals – Copernicus was born to a rich merchant in 1473; Tycho Brahe was born the son of a nobleman in 1546; Galileo was born to an educated family in 1564 – but it widened to an army of enthusiasts, amateurs and innovators from all classes. To join this community, all you needed was a succession of clear night skies, some enthusiasm for mathematics, access to other people's books and papers, and some good instruments. If you didn't like the telescope you started with, then it was up to you to build a bigger, better one. If you didn't trust someone's calculations, and didn't believe their theorising, you could do your own and then write a letter or publish a paper explaining why.

Out of this network of the obsessive, the impassioned and the eccentric, astonishing things emerged. Galileo tried to discover whether light had a speed, or was transmitted instantaneously; and then he set out to establish whether heavy things really fell faster than light ones; he played with lenses, to observe small things on Earth and see mountains on the Moon and spots on the Sun – and because he devised experiments and recorded his results and drew conclusions from them, he is heralded as one of the founders of modern science. To make sense of what he saw through his telescope, Isaac Newton devised a theory of optics, and provided a new tool for all science; he formulated the laws of motion and composed a theory of gravitation, and thus ushered in the space age; he devised a new mathematics; and when he died, to be buried in Westminster Abbey, Voltaire remarked that England had honoured a mathematician as other nations honoured a king.

But even before Newton had been born, Jeremiah Horrocks, an impoverished young Puritan from what is now Toxteth in Liverpool, had read the works of Kepler and Brahe, calculated that the Moon's orbit was an ellipse, predicted and observed the transit of Venus across the face of the Sun, and made a stab at estimating the distance between the Earth and the Sun, all before his death at twenty-two. Horrocks was just an early member of the army of the impassioned, the star-struck and the self-taught, the university-educated, the navy-trained and the government-funded, the hobbyists, society members and loners who for the next four hundred years got to know, intimately and with the naked eye, every one of the thousands of stars in the night sky, who spotted comets, tracked asteroids and observed eclipses, counted sunspots, built ever more sophisticated telescopes and proposed ways of calculating the distances to the furthest stars. They, and their increasingly well-equipped colleagues in the observatories and research laboratories, shared information across space and through time, building up an unparalleled corpus of data, ideas, solutions, hypotheses, experiments and surveys. Nations might war with each other, and kings dispatch armies to foreign territories, but the cooperation between astronomers hardly faltered, and Albert Einstein's theory of special relativity – proposed in Germany before the First World War – was promptly tested after that bloody and destructive episode by a team of British astronomers led by Sir Arthur Eddington. Astronomers by then were a professional class: they had access to increasingly expensive telescopes and an ever-widening range of spectral wavelengths, but they drew support and encouragement and recruits from a huge army of amateurs, and most of the comets identified in the twentieth century were recorded by enthusiasts rather than professionals. This community of the starstruck, in four hundred years, began to create an atlas of the heavens and a taxonomy of the heavenly bodies.

They identified new stars – novae – and then spectacular bursts in the skies called supernovae; they identified clouds of gas called nebulae and eventually pinpointed the bright new stars that form within these stellar nurseries; they saw white dwarfs smaller than some of the planets, and red giants and orange giants, and even supergiants, that had swollen to circumferences far greater than the orbit of the Earth around the Sun. They identified very bright stars a million times more luminous than the Sun (but too far away to be seen at all with the naked eye), and very dim stars with a 20,000th the luminosity of the Sun. They used parallax techniques to calculate the distance of the nearer stars, by measuring the angle of observation from different positions along the planet's orbit. They also began to use brightness as a proxy for distance – an uncertain measuring rod, given the natural variations in luminosity. But what gave them increasing certainty about the answers was the sheer number of attempts to get the technique right, and the enormous network of discussion, comparison, debate and disagreement, not just around the planet, but through time as well. Had a new idea, a new technique proved unsatisfactory? Then you could go back to a bright idea abandoned during the last century and see whether it could be used in a different way. Was the new approach inherently unsatisfactory? Or just an answer awaiting questions that had yet to be posed?

Behind the observers who prayed for dry, clear skies and dark nights were teams of theorists who followed the logic of observation. Long before Stephen Hawking, theorists had calculated the stupefying mass of white dwarfs and conjectured the possibility of black holes. Alongside these people were scientists who had begun to realise that light itself contained coded secrets: if you analysed its spectral content you could begin to make guesses about the chemistry of its origin. Helium was observed during an eclipse of the Sun in 1868, but not found on Earth until 1895. Using spectroscopic analysis, these secondary observers began to understand what stars were made of, and then ultimately to realise that they were the cosmic forges that began as hydrogen and then, in the course of their billion-year blazes, fashioned the other ninety-one elements. Yet other observers realised that they could work in the daytime: they could tune in to radio signals not just from stars but from clouds of gas between the stars, to detect phenomena they could not see, and discern the universe in a different dimension. The space age took astronomy to new heights: above the filter of the atmosphere, space-borne cameras could measure the universe in the infrared and ultra-violet; they could collect x-rays and gamma rays from hectic, destructive events in galaxies far away and long ago, and they could tune in to the whispers of creation. They puzzled over pulsars and magnetars and quasars, and they mapped and named variable stars – Cepheid variables, for instance, that gleamed and faded and gleamed again, because they heaved and sighed rhythmically in a slow death by stellar exhaustion.

One researcher, in 1912, discovered something immensely practical about Cepheid variables: the longer the periods between the bursts of brightness, the greater the luminosity of the star. It became possible to measure this luminosity, and discover that it could match the brightness of a hundred, or a thousand, or ten thousand times the Sun. With this discovery, astronomers had a yardstick for measuring immense distances, and it was the light from the Cepheid variables in Andromeda that revealed it to be not a dust cloud in the Milky Way, but a separate, faraway galaxy.

There were other kinds of variable: stars like Algol that seemed to shine and then weaken every 2.9 days, evidence of a little waltz being performed by a pair of stars – one bright, the other massive but too faint to detect. The new astronomers looked not just at the stars, but at the spaces between them, and then they detected, in the darkness between the galaxies, the faintest radiant echoes of the colossal big bang that launched the visible universe 13.7 billion years ago, and then much later they managed to trace the tiniest fluctuations in this cosmic back-ground radiation: the minuscule initial differences in the smoothness of the energy of the nascent universe that must have permitted matter to clump, and stars to grow, and galaxies to foregather. They identified colossal black holes – some of them invisible monsters that had already consumed the mass of a billion stars – at the hearts of distant galaxies, and from that concluded that there could be one at the core of the Milky Way, invisible not just because that is the nature of a black hole, but undetectable because it was screened by the millions upon millions of stars between us and the galactic heart.

They also, in the last century, discovered one of the galaxy's great and enduring secrets: most of it is made of matter that can be neither seen nor felt nor identified. This dark matter revealed itself at first as a kind of gap in the arithmetic of the cosmos. Galaxies are creatures of stars and dust and gas: with powerful telescopes you could stare at one and begin to make very crude estimates of the number of stars and their average mass. That is one way of guessing the mass of a galaxy: add up what you can see and make a total. But there is another way. You could use the rotational velocity of the distant stars to make a crude estimate of the gravitational force that held a galaxy together, and from that calculate the mass associated with that gravitational force. But whenever people tried both approaches, they discovered a mismatch: there was far more gravitational force than could be explained by the mass of visible stars.

The latest guess is that the myriad bright lights of the heavens are just a fraction of all that there is. Between the living stars and the dead, the gas of the stars yet to form and the dust of stars that exploded long ago, there is another form of cosmic building-stuff, a strange, tenuous, undetectable dark substance that makes up more than four-fifths of everything in a galaxy. It emits no light. Other material passes through it without glancing, bouncing or slowing. But it is there, and reveals itself in its subtle, invisible gravitational grip.

There are individual heroes in this story, and others, men and women who put their names to papers that made scientific history, but their achievements were built on the evidence, the conjecture, the inspiration and the ingenuity of an army of observers and theorists, a disorderly community of individuals and small groups united by a passion for the night sky. Astronomers and their colleagues in the physical sciences were the first to make sense of the substance of our flesh and the light in our eyes. And in the course of this disorderly hubbub of curiosity, cooperation and competition, the community of heaven-watchers also began to assemble, slowly and with a great deal of uncertainty, a geography first of the galaxy, and then in the last century of the universe itself.

It is not a complete geography. The British broadcaster and astronomer Patrick Moore once compared our observation of the heavens to a man standing in Piccadilly Circus and trying to work out the shape of London. The Milky Way, that smear of faint light that most urban dwellers would be lucky to see at all, is actually a barrier that screens a substantial chunk of the universe from view. We can see it, but not through it. So the heart of our own galaxy is concealed from us, and, of course, so are all the stars on the far side. Even so, the community of astronomers has collectively observed other galaxies, and compared them to what little they could see of this one, and collectively agreed that the Milky Way is a barred spiral galaxy, containing a mix of stars, some almost as old as the universe itself, some still in the act of formation from the primordial dust and gas, all in a swirling structure 100,000 light years from rim to rim, and a thousand light years from top to bottom, made up of a massive, bulging centre and trailing from this at least four spiral arms, swirling rather as strands of rope or twine might swirl from a spinning disc. Our Sun and its planets are about 30,000 light years from the galactic centre, on the edge of one of four spiral arms, known as the Sagittarius arm. Whether the galaxy truly has four arms, rather than six, or two, and whether it really is a barred spiral galaxy, with a bar of ancient red stars running through its centre, is something we cannot know for certain: our neighbours in Andromeda or Draco or the Large Magellanic Cloud will have a clearer idea of our status in the galactic zoo. But for the time being, this community has found an address for us: the solar system, near Sirius and Barnard's Star and Alpha Centauri, the Sagittarius arm, the Milky Way, the Local Group, the Virgo Supercluster. We have our place in the sun, and our Sun has its place among the stars.

Not forever: as we know, Andromeda is heading our way. Andromeda is 2.3 million light years from the Milky Way, but the gap between the two galaxies is closing at the rate of 140 kilometres a second, or 270,000 miles an hour, or 6.5 million miles every day. This pace is not easy to imagine, but set against the speed of light, it is glacially slow, and the projected collision date not only leaves modern humans time to set their affairs in order; it also leave them time to colonise the rest of the galaxy, or to become extinct altogether, leaving no issue. The moment of expected impact, too, incorporates a number of uncertainties. The rate of approach is one of them, the distance to Andromeda is another. The gas, dust and other stuff in between could exert a drag and slow the pace of events. And the approaching monster is not only falling towards us; it is also moving sideways at an unknown speed, and – supposing this speed was fast enough – could miss the Milky Way altogether. But T.J. Cox and Abraham Loeb of the Harvard-Smithsonian Center for Astrophysics in Cambridge, Massachusetts, were, in 2007, fairly certain that the two cosmic communities would be in close touch; certain enough to start drawing up preliminary plans for the redevelopment of the neighbourhood.

One of these is that the two heavenly bodies will fall into a slow gravitational embrace and waltz around each other for an indeterminate time, bumping against each other a couple of times before finally merging. Since each of the galaxies is spinning like a discus as it flies across the emptiness of space, there is a chance that any remaining observers in the solar system won't get to see the collision at all: they could be on the safe side, their view obscured by the density of stars at the heart of the Milky Way. But there is also a chance that they will be first-hand witnesses to a celestial takeover bid. They calculate that on the first brush with the big bruiser – Andromeda is perhaps three times more massive than the Milky Way – there is a 12 per cent chance that the Sun will be dragged from its place on the Sagittarius arm, and sent spinning to the farthest fringes of the galaxy, to end up as a member of a kind of ribbon development of stellar neighbourhoods, part of a semi-detached tail or wisp of stars drawn from the Milky Way by the attraction of its muscular neighbour, washed away by a gravitational tide. That's just the first brush. After the second bump and grind between the two bodies, there would be a 30 per cent chance that the Sun would be relegated to a permanent place on the fringes of the stellar megalopolis. But there is also a tiny chance – 2.7 per cent – that the Sun will become detached from the Milky Way and become part of Andromeda before the dancing stops, the merger is finally completed, and the two galaxies become one.

What any notional observers a few billion years from now will actually see is anybody's guess. Head-on collisions between the opposing stars are unlikely. That is because, while the stars are huge, the spaces between them are so much greater that the encounter would be rather like the meeting of two blasts of birdshot at a clay-pigeon shoot: the pellets are far more likely just to sail past each other. The stars, being so massive, would certainly exert an effect on their nearest neighbours as they rushed closer to each other; they might also rip material from each other, and – since they are numbered in each galaxy in their hundreds of billions – some of them would certainly collide. But any eyewitnesses in the new, semi-detached, faraway solar system would be very lucky to see such things. As usual, if there is any action, it will be downtown, among the bright lights and hot venues at the hearts of the newly-merged galaxies.

ELEVEN

The Universe

All There is

We are all lost in space: we none of us know where we are. That is because there are no coordinates in the cosmos at large. The Copernican principle, the one that says that we are not the centre of the universe, and that there is nothing special about our place or our time in the great scheme of things, is one of the foundation stones of modern science. The revolution launched by Copernicus four hundred years ago tells us that we are not the focus of creation, a little cockpit of imperfection enclosed by the heavenly order of the celestial spheres. It tells us that we are on a fortunate but not-out-of-the-ordinary planet orbiting a nondescript main sequence star nowhere very special on the outer limb of a neither-here-nor-there galaxy in an undistinguished corner of the universe.

Put like that, the Copernican principle seems to say that we should know our place in the great scheme of things: that wherever it is, there is nothing very special about it. In fact, and perhaps for that reason, there is no map that we could ever devise to pinpoint it. Earthbound astronomers use a grid system first devised by the Babylonians to look out to the stars. This grid system cannot describe reality, because it *does* assume that we are at the centre of the universe: of course it does. We are here, and everything else is out there. Moreover, it only serves as a guide to the immediate vault of the night sky and its most visible stars, and it would be no help if we left the galaxy and accelerated away into the cosmic emptiness that is the space in which all the galaxies are embedded, like little oases of light in the never-ending blackness.

Cosmologists are fairly confident that they have the measure of the density of the universe: it adds up to an atom or two of hydrogen per five cubic metres of space. But almost all of this hydrogen is clustered, glued or crushed into stars and galaxies, leaving the space between the galaxies as empty as empty space could be without being a real vacuum. There is something – quintessence, dark energy, antigravity, the almost meaningless labels are themselves a testament to how little we know about this stuff – that holds space together, makes it stretch and distort, inflate and expand, respond to matter and control the speed at which matter accelerates. So space is a thing, a fabric, a surface, an entity. It is not *nothing*. Nothing is the region from which space emerged, just as eternity is the backcloth from behind which time first took a bow. So space, or to get pernickety, space time, is real. It is a here-and-now, with a much more compact there-and-then, and a limitless and ever-expanding future ahead of it. Space is a vast horizon with direction and order and billions of way stations made up of clusters of billions of blazing stars in every direction, all of them different, and differently grouped, and in different stages of evolution, but – on the biggest scale – all of them pretty much the same. Look one way, and you see galaxies and clusters of galaxies, filaments of super-clusters of galaxies and huge voids that seem to contain nothing. Turn the other way, and you see pretty much the same distribution of light and dark, of matter and emptiness. This sameness is one of creation's great puzzles: how did an expanding universe 'know' that it had to be much the same in every direction?

It would be an even bigger puzzle for the captain of a hypothetical *Star Trek* ship that really could travel far faster than light, as this sameness would make it very easy to get lost in space. Because, on the grandest scale, everything is the same, there are no signposts in the cosmos, no highways, no direction home: there is no map, no diagram, no graphic aid that makes any sense to a navigator. There may never be one. It took a Columbus, and a Magellan, and a Captain Cook to begin ordering the continents and islands on the surface of planet Earth, but no cosmic explorer could ever divide the universe into sectors or quadrants or destinations. There is no universal meridian, no centre or rim, no time zones, no datelines, station concourse, no waiting room. In intergalactic space, you could be anywhere, or nowhere, or everywhere. There is nowhere to rest, and nowhere at rest. We are trapped in a universe that has a beginning but no starting point; a universe that is finite but has no end.

Once I have left my galaxy – left it far enough behind to no longer feel its gravitational tug, left it far enough behind to see it properly, as one of a local group of galaxies, a cluster of stars in kind of multi-galactic province – there is no way back. For one thing, I do not know what my galaxy looks like. No long-term resident of the Milky Way really has any way of identifying the neighbourhood from beyond its borders. There are representations of the Milky Way galaxy, but they are inferred from what little we can see, and these inferences are then matched against the varieties of galaxy we seem to see across huge distances. From within, the Milky Way looks like a spiral galaxy, 100,000 light years across and a few thousand light years deep. But that describes seven out of ten galaxies seen so far. And we only *think* the Milky Way looks like that: we have no passport photograph, no police mugshot, no objective image that we could use to identify this larger environment in which, somewhere, is a star that we call ours.

It is as if a man who had spent his entire life in a living room with a window looking onto a small, fenced garden was hastily abducted, blindfolded, driven an indeterminate distance and then released in a large housing estate and left to identify the domicile that had sheltered him for so long. Although he might have a fair idea of what other houses look like, by analysing what little he can see of their structures through the only window available to him, he has no evidence that his own house is designed in the same way, or has any of the features that he could see from his domestic prison.

Myth and maritime experience suggest two ways to help the voyager get back to where he belongs. One of them is the example of Ariadne's thread: when Theseus went into the Labyrinth to challenge the Minotaur, he laid a trail of silk as he went, so that he could follow the path back. But let us abandon the idea of a clew. There is no substance, however fine, that a traveller could spin behind him in a journey measured not just in light years, but in millions of light years. The other is dead reckoning: you take a careful note of the direction of your journey, and the speed at which you travel. That will tell you, at the end of a set time, roughly where you are. To get back, just make a 180-degree turn and head home, being careful to allow for variables such as wind and current and the progress across the sky of the Sun and the stars, supposing you had used them as directional indicators. Dead reckoning is a haphazard technique in a featureless desert, and pretty tricky in the open ocean. It offers no security for the space traveller.

There is a special problem for the entirely notional *Star Trek* voyager, the happy imaginary adventurer who can travel at speeds far faster than light; who can be a hundred million light years from home in the very next episode. Suppose I really could get a hundred million light years from my home galaxy in a week, then I would have no way of identifying my starting point anyway, because the light from it would have taken a hundred million years to get to where I am now: that is, I would be looking back in time, to a galaxy that contained a star around which circled a planet on which the Atlantic Ocean did not exist, and on which even Tyrannosaurus rex had yet to evolve. So my lodestar would not be the one from which I set out a week earlier, and could not become my home for another hundred million years. But there are other, more practical reasons why I would be lost in space. The galaxy that I have left behind – that is, escaped its gravitational field – is also leaving me. It is simultaneously rotating upon a central axis and falling towards another galaxy. And the space between the galaxies is expanding. How fast it is expanding I cannot say, because velocities such as miles per hour or kilometres per second begin, soon enough, to have no meaning. The rule of thumb seems to be that the further away a galaxy might be, the faster it is receding, and since the universe is so vast, the furthest galaxies are receding at something approaching the speed of light. This, too, is a bewildering statement, because in a vacuum, light travels at a fixed speed. Light that radiates from a star that is hurtling away from us still arrives at the speed of light, but the wavelengths of this light will betray this recession because the light that we see will seem shifted towards the red end of the spectrum. This is called redshift. It provides a measuring stick for cosmologists. If the light from the most distant galaxy – let us say it is twelve billion light years away – has a redshift that indicates that it is receding at 75 per cent of the speed of light, then it must have been receding at that speed when the light I now see left it twelve billion years ago. In which case, where is it now? And if I do not know where my landmark is now, how could I know where I myself am, now or at any time?

The confusion starts right at the beginning. That there was a beginning, no one now doubts. If all the galaxies are receding from each other at ever greater speeds, then there must have been times in the past when they were ever closer together. At some point in the past – consensus puts this point at about 13.7 billion years ago – everything in the universe was crammed into a volume far smaller than the nucleus of an atom. This is ever so small, but only if you imagine the nascent universe from the outside, so to speak: from the region into which the universe began to expand, before it expanded. But there was no 'outside', just as there was no 'before': space and time emerged together, locked in some inseparable embrace. St Augustine of Hippo figured this out for himself, and gave his reasoning in *The City of God*, the book he was writing as the Western Roman Empire fell to the Vandals. If God had created the world in time, then you'd be tempted to ask: what was God doing before He created the universe? And if He created the universe in an already existing space, then you'd feel impelled to ask why He made the universe here, rather than over there. 'We are not to think about infinite time before the world, any more than about infinite space outside it. As there was no time before it, so there is no space outside it,' St Augustine argues. If St Augustine, Stephen Hawking and almost everybody else who has done the sums is right, then from the start, the universe contained everything and everywhere even when it was only a fraction of the volume of the nucleus of an atom, and under those circumstances it still would have been pretty easy to get lost in: a world crammed with infinite possibilities stays the same, whatever its dimensions.

But we don't have to think of someone 'lost' in the very early universe. There is nothing that we could recognise, identify, pinpoint or mark in this cosmic fireball from which everything we know and see and touch has emerged. The early universe is a strange place, but its evolution is confidently calculated by cosmologists in dimensions measured not in metres but in fractions of a second of time expired: fractions against which a thousandth of a second would seem like an eternity, and even a billionth of a second would drag on like a geological era. If you press cosmologists and particle physicists on questions of dimension immediately – that is, an unimaginably small tick of the cosmic clock – after the 'big bang' they rather reluctantly talk about the universe when it was the size of a grapefruit, or a basketball, or the dome of St Paul's; in their minds it has already undergone a condition called cosmic inflation, and has already expanded, far faster than the speed of light, from virtually nothing to become something hot and dense and bursting with all that was, is and ever will be.

Why are they reluctant to provide dimension, and a sliding scale of growth, a geography and a list of original contents for this rapidly expanding empire of everything? In the first place, because it doesn't make much sense to talk about the unknowable, and in the second because it doesn't make much sense to talk about it in words.

The language of creation is probably entirely mathematical: in the confined world of relativity theory, quantum mechanics and other refinements, the tiny universe contained in the first billionth of a second of time starts to make sense. Things that are mathematically impossible are excluded. Things that are not excluded by mathematical reasoning are permissible, possible, probable or perhaps inevitable. In mathematical physics, the not-impossible happens not just every day, but every second. Even where there is nothing – neither stars nor space nor time – 'virtual particles' of ridiculously small dimensions can borrow energy from the vacuum, that is, from nothing, and wink into and out of existence in preposterously small fractions of a second, paying back their borrowed energy as they go. That's what the mathematical reasoning says. If a fragment of matter, however small, can exist for a period of time, however short, on borrowed energy, then it follows that a universe of two hundred billion galaxies, each containing two hundred billion stars, can exist for billions of years, with capital borrowed from the same lending market. Think of the universe as a property purchased with a 100 per cent mortgage: value and debt cancel each other, and as proud owner of not one but two huge responsibilities, you are worth precisely nothing. Does that bother you? Not if you can safely call the house yours for the extent of your life-time, and hope to hand it on to the next generation. In the case of the universe, the capital – the energy to expand – is matched by the collateral implied in the gravitational field, or negative energy. So the total energy of the universe, huge as it is and stuffed with stars beyond number, is zero.

Having established that a whole universe – space, time, stars and stardust stretching to a horizon a great deal more than 13.7 billion light years away – may exist independently of any Creator, first cause or Prime Mover, mathematical physicists can begin to work backwards from what is, to what must have been. In this ultimate detective story, when cosmologists confer, they talk of things that cannot be seen and may never be found, but that have an existence somewhere, if only as a hypothesis: magnetic monopoles, for instance, and cosmic string,* superstring, seven extra dimensions, branes, gravitons, gluons, Higgs bosons,

---

*Depending on which physicist you talk to, magnetic monopoles would either have no dimension at all – a point in space with no measurements – or they would be 100,000 billionth the size of a hydrogen nucleus, which is pretty small. They shouldn't be too hard to spot, because they would have the mass of ten billion billion hydrogen nuclei. Cosmic string is another hypothetical fabrication of the very early universe. It is described as a kind of a topological fault in space time. It has just one dimension: length, so cosmic string could pass through an atom without disturbing it. Cosmic strings come in huge loops or threads measured in millions of light years, they move incredibly fast, and they are massive: one inch of cosmic string would, the theorists say, weigh ten million billion tons. A mile of cosmic string would match the mass of planet Earth. One cosmic string theorist has argued that these awesome filaments of nothing might have provided the gravitational thread along which all the galaxies are now strung like pearls. Nobody, however, has so far with much confidence identified any of this cosmic old rope. Alan Guth of the Massachusetts Institute of Technology and the author of *The Inflationary Universe: The Quest for a New Theory of Cosmic Origins* (1998), is the man who first proposed that the universe was the ultimate free lunch, served up by cosmic inflation.

supersymmetry, weakly interacting massive particles, vacuum energy, Omega, singularities, quantum fluctuations, scalar fields, de Sitter space, Misner space, Calabi-Yau space, Cauchy horizons, closed timelike curves, false or metastable vacuum, Einstein-Rosen bridges, energy-momentum tensor, quantum tunnelling, baby universes, wormholes.

Anything that can be predicted must then be identified: if it is there, it confirms the prediction, and if it is not there, its absence must be explained. Some non-presences are more easily explained than others: their existence was predicted, and predicted to be fleeting. So, strange animals that emerged in the first trillion trillionth of a second of time would have disappeared with equal promptness, decaying into a succession of other exotica, cascading towards the kind of stability that – by the end of the first few minutes – would result in the raw components that would finally result in atoms, dark matter and dark energy. Many of these first strange apparitions could only be identified from the footprints left by their descendants, many generations later, always supposing you knew what the footprint should look like, and what kind of lineage to contemplate. Others should still be present, 13.7 billion years on, very massive and impossible to miss.

One of these, the magnetic monopole, was predicted so convincingly, and assigned such an enormous mass, that it precipitated one of the great ideas of cosmology: cosmic inflation. Suppose – conjectured a young doctoral student set the challenge of explaining its absence – the universe had started very small, but then inflated itself so fast and on such a vast scale that the initial large and very-hard-to-miss delivery of magnetic monopoles became hopelessly diluted by the immensity of space. This period of inflation began and ended long before the first billion trillion trillionth of a second had elapsed, but by the time it ended, magnetic monopoles would be as hard to find as a handful of currants in a bun puffed up to the size of the solar system.

Conjectures such as these make no sense if you try to imagine them, to give them a tangible, palpable substance and dimension. They do make sense within mathematical physics. That doesn't mean they are true; it means they are consistent with mathematical logic. But this logic, however improbable, seems to conjure up reality. A mathematical physicist in 1928 predicted that matter would be matched by a thing called antimatter; by 1931, other scientists had spotted antimatter. Now particle physicists routinely manufacture antimatter in giant accelerators. In 1948, mathematical physicists proposed that if the cosmos was expanding, it must be cooling, and that however cool it had become, even the vast emptiness of space between the galaxies would still be glowing feebly with the fading heat of the first great cosmic explosion. The cosmic background radiation they predicted was identified by radio astronomers in 1965. Plato observed, twenty-five centuries ago, that the gods must have practised geometry. And, Plato's inheritors would now add, quantum electrodynamics.

All of this is to explain why, when he answered the question of why there were no magnetic monopoles to be seen, Alan Guth didn't burst out laughing and tear up his calculations. It was a logical answer, not *the* answer. It was only months afterwards, after talking to other physicists, that he realised cosmic inflation would explain several simultaneous puzzles about the universe, one of which was why it was the same in every direction, and that therefore it might offer a real explanation, rather than simply a logical one. Cosmic inflation is now an orthodox cosmological faith, a broad church with more than a hundred versions competing for sectarian supremacy. It has made predictions – most notably that there would be tiny variations in the cosmic background radiation, measured in hundreds of thousandths of a per cent – and these too have been confirmed.

This doesn't mean that cosmic inflation is true, or is the only explanation; but so far it offers the best explanation for why we are all lost in space, trapped in a universe that is at the same time both explicable and enigmatic, and going nowhere fast. I believe that I am sitting still, in a small room in a small town in East Sussex, England, but I am travelling towards the night at a thousand kilometres an hour as the planet rotates, and the planet, too, is travelling around the Sun at thirty kilometres a second. Our parent star, in its turn, is rotating around the heart of the Milky Way galaxy at two hundred kilometres a second, and the galaxy itself is crashing towards Andromeda at 120 or more kilometres a second, but also rushing away from the most distant galaxies at nearly 200,000 kilometres a second. The furthest galaxies are expanding away from each other at speeds that are getting close to the speed of light. The implication is that beyond these galaxies, far beyond any imaginable horizon, there might be galaxies that have, from our point of view, already exceeded the speed of light. This in turn means that as far as they are concerned, we here on Earth are receding from them faster than light. So, in a sense, cosmic inflation is still going on, somewhere else in the universe. It stopped, but only in our part of the cosmos. But if cosmic inflation happened here, once, why couldn't it happen again, somewhere else?

The word 'universe' implies one big place, but the logic of cosmic inflation suggests something different. If our universe could inflate itself up from nothing, why should it not happen again, and again? Could this universe be just one of an infinite foam of universes, bubbling out of the bottleneck of eternity? So, contrary to the rules formulated by both St Augustine and Stephen Hawking, cosmologists have started talking about 'before' the big bang. Can a universe really conjure itself from absolutely nothing, and if so, why should ours be the only one that did it? Are we the consequence of an accidental episode of cosmic inflation that began in someone else's universe, a so-far-successful but temporary bud from the tree of eternity? Are we just at home in a benign region of a universe that might look completely different on the far side of the horizon, which of course we could never reach? Do we occupy a multiverse, in which all sorts of conceivable and difficult-to-imagine versions of reality exist simultaneously, only one of which we think we inhabit? Is there something about the physical properties of this universe that requires that life would inevitably emerge, but only after a couple of generations of galaxy formation and decay? And if so, docs that mean that there must be billions of other universes that wink into existence, survive for minutes, hours or millennia, and then collapse back into nothing because the physical properties issued at the moment of universal random generation didn't quite make sense, didn't quite add up to stars, galaxies, liquid water, ninety-two elements and the right kind of biochemical soup?

There are even more disconcerting questions: if there was a beginning, will there be an end, and what form might it take? The logic of inflation theory – that a universe can conjure itself from nothing, that it can 'borrow' mass and energy from nowhere, that it can secure a 100 per cent mortgage on itself without further collateral and without interest or capital repayments – also carries with it the idea of eventual foreclosure. Forty years ago, after the first confirmation of the 'big bang' hypothesis, it was possible and even fruitful to speculate along theological lines: some Thing or some One lit the blue touchpaper, the universe was created in a cosmic explosion and gradually but inevitably took shape, along with the vault of heaven, the stars, the waters and the dry land, and eventually the creatures upon it. It didn't happen in seven days, but it happened in sequence, over time, and it would of course be a temporary event. The universe was expanding, had expanded, would gradually slow its expansion, grind to a halt and then start to collapse in upon itself. It would end, as it had begun, in fire. This outline contained within it something for everybody. People who wanted to believe in a First Cause, a Prime Mover or an old-fashioned God could accept this story as evidence that God moved in mysterious ways his wonders to perform, and it left a role for God. There is no obvious reason why so many of the physical properties of the universe are tuned so exquisitely that they permit the existence of galaxies, stars, planets, biochemical soup and finally life and then – after a very long time – intelligent life: OK, that was God's choice, and we are here to celebrate His wisdom. Unfortunately, the same story was also perfect for people who wanted to believe that there was no God: the universe began out of a singularity and would collapse back into one. Who is to say a new version of the universe would not rise from this new collapse? And if so, who is to say that our universe was not simply a recycled form of a previous one, and so on, ad infinitum, a Latin tag that, for once, really is used correctly?

This version also provided huge entertainment for all kinds of storytellers. If the universe ran backwards, would time and space run backwards too? Would the dead rise from their graves, get younger, cough up apples, put them back on the tree, lose their memories, regain their innocence and eventually – in some sort of Freud-made-flesh apotheosis – return to the womb? Would we get a second chance to see it all again, this time backwards, and if so, would we know this? If space and time reversed themselves, then perhaps our thought processes would run backwards too. In which case, how did we know the universe was expanding now: perhaps it had already begun to contract? Time's arrow, shot vertically into the vacuum, might fall back feathers first; at any single instant on its downward path, it would still look as though it were pointing the other way, because that is the way we see arrows, with the sharp end pointing somewhere, and not away from it.

Cosmic inflation changed the big picture, and recent cosmological research and astrophysical observation have opened up a huge new range of questions. For instance, it no longer looks as though the universe will necessarily slow its expansion, run down and collapse back upon itself. Perhaps it could go on expanding forever. It would get bigger and cooler, and the galaxies would grow further and further apart. Old stars would go out, new stars would be increasingly less likely to form. After 150 billion years, the Milky Way galaxy would be alone in its cosmos: all the other galaxies would either have merged with it or fled across the light horizon, leaving only a backcloth of the blackest empty space. Finally, trillions of years from now, the universe would still exist, and everything in it would be at the same temperature of almost absolute zero, and all life, all exchange, all motion would have come to an end, but space would continue to expand, and time would continue to tick away, forever. Since time emerged from eternity, that raises the disconcerting prospect of two parallel eternities, one of them infinitely shorter than the other, since it had a beginning, while the other had neither a beginning nor an end.* This scenario is called the heat death of the universe. It would contain no witnesses, no illumi-

---

*There is always a temptation to quote Douglas Adams on the size of the universe: the bit from *The Hitchhiker's Guide to the Galaxy* about how mind-bogglingly big the universe is. The really mind-boggling thing is that, 13.7 billion years after the event, people can confidently sit down and calculate what must have happened, make predictions and then run little experiments that show that they are on the right track. Einstein said the most incomprehensible thing about the universe was that it was comprehensible. Steven Weinberg, in a book called *The First Three Minutes* (1977), said: 'The more the universe seems comprehensible, the more it also seems pointless.' Paul Davies, in his book *The Goldilocks Enigma: Why is the Universe Just Right for Life?* (2006), wonders what would happen if – having just got the hang of cosmic physics and calculated the history of everything with some confidence back to the first thousandth of a second of time – the human race was extinguished in some advancing cataclysm. 'It may never happen again,' says Davies. 'The universe may endure for a trillion years shrouded in total mystery, save for a fleeting pulse of enlightenment around one average star in one unexceptional galaxy, 13.7 billion years after it all began.'

nation, no information, no means of communication, no postal address anywhere within it: all places would be alike, dark and cold and alone. It would be a black whole, rather than a black hole.

Or would it? Are diamonds durable? Will hardware always hang around? One set of theoretical calculations suggests that even atoms themselves have a limited life. After a billion trillion trillion years, perhaps protons and neutrons and the dark matter that surrounds them will decay, burst into tiny bubbles of radiation, and light up the dead, cold universe in one last slowly flickering firework display. After that, the universe will endure forever: time will tick away, and space will continue to expand, and light will continue to spread across the ever-increasing cosmos, but there will be no one to see it, and indeed nothing to see, as the space between individual photons continues to grow. Or perhaps the dark energy that – for the moment – supports the accelerating expansion of the cosmos will get tired, run out of steam, give up, and permit a massive, unexpected collapse. Or perhaps the huge confidence trick on which the universe exists – you borrowed the world, one day you will pay it all back – will falter, and the fabric of space time itself will rip apart: under this scenario there could be no warning, no slowing down and running backwards at an ever-accelerating rate towards a big crunch. The universe would dissolve at the speed of light. In the long run, the forecast is unpromising: in the long run, we are all dead and everything around us is as good as dead too.

The good news is that we have a very long run ahead, and a crowded universe to explore – providing, of course, that we can find our way around in it. That would require a reliable map, but I cannot think of any map that would remain accurate for long enough to be of any use to some entirely hypothetical faster-than-light universal circumnavigator. The first European mariners to attempt to map the Pacific had a hard enough job finding the archipelagos and atolls that they already knew existed – and when they did find them, sometimes thought they were discovering entirely new islands, because coordinates like latitude and longitude were so hard to establish precisely, and dead reckoning so perilous a navigational aid. But if the landfalls seemed to duck and weave, and steal away to the wrong locations, at least the Pacific remained the same overall size. What kind of homecoming could there be for the lonely long-distance runner in the cosmos, when every solid, visible, gleaming galaxy is on the move, either receding ever faster with distance, or falling headlong towards some other galaxy, and the fabric of space itself, in the grip of an unknown and perhaps unknowable daemonic force called dark energy, is stretching apart as you try to cross it? Indeed, what kind of universe would it look like if you were travelling faster than the starlight you had hoped to navigate by, searching for your own place in the Sun whose rays would – from your point of view – have ceased to shine? And what might befall such a voyager on the journey into the unknown? This universe must be vastly bigger than the corner we can observe, but perhaps at some point it opens onto another universe, and yet another. Any such journey is a one-way trip. There is no way back.

And, of course, we must already be on this journey anyway: we may think we are going nowhere, but from somebody else's point of view, we must be receding into the distance at a measurable proportion of the speed of light. Anybody over there who sees this galaxy, this star, the reflected light from this planet, must in fact be seeing what we were billions of years ago, because we are separated by distances measured in billions of light years. So, no letter from over there could ever reach over here, and a wistful correspondent on the far side could never have intended it to have been for us, the present tenants. And if, somehow, one day, such a letter did arrive, we too would have packed up our planet and moved on, leaving no forwarding address, and nobody to scribble on the envelope that laconic counsel of despair: 'Return to sender.'

At a very young age, I once tried to imagine space travel: tried to imagine leaving the solar system and the galaxy and the travelling through space until I reached the end of the universe. (If, as the Sisters of Mercy who had then taken charge of my primary education were correct, and the universe had been formed by a Creator who had always been, then He was outside the universe, so there had to be an end or a boundary.) So what was at the end of the universe, the final enclosure of space? My seven- or eight-year-old self concluded that space flight would ultimately be arrested by something viscous and increasingly impenetrable – a substance close to the consistency of Vicks VapoRub, a product parents used to rub on the chests of children with coughs and colds – through which there could be no passage, and therefore no discovery, and therefore no more space. It wasn't much of a solution, but it was an early personal examination of a problem that has on occasion perplexed the greatest minds. But if the universe is just one functional living space, that has budded from a pre-existing universe, that co-exists with all the other universes that have ever been or will be – if we live in a multiverse – then the problem takes a different shape. Once lost in space, I have no need to go home. My multiverse is the equivalent of an infinite universe. If the laws of physics are the same, then in an infinite universe, anything that can happen will happen, and happen an infinite number of times. So in the infinity of very different possible worlds, there will also be worlds that are quite like, very like and exactly like the one in which I first wrote, in an exercise book, the name of a street, and the town, and the province, and the country, and the globe, and the galaxy. So I just carry on (this is a thought experiment, but even in a thought experiment, it is going to be a preposterously long journey) surveying all the solar systems with Earthlike planets, and then all the Earthlike planets with Moonlike moons and continental structures that look like Eurasia and the United Kingdom, and then all the United Kingdom lookalikes that have inhabitants that look remotely like the British, some of whom look like people I used to know, until I find someone who looks like me, and bears my name again, and sits at a desk like mine, surrounded by books all of which I recognise.

Which is where we came in: I am once more in a room, the books are the same, the desk is the same, but now the room is plaster-walled and painted white, the bookshelves are more orderly, and the view from the window is different. I still face south, but now I can see other houses, and beyond them, on higher ground, more distant trees. We have moved. We have begun again, and in a new town, in the same county. I have made a new entry in my address book. I am here, in a palpable address with a precise postal code and exact geographical coordinates. But friends who visit still telephone first and say, 'Where exactly are you?' and the question still has an odd, unsettling effect, as if my answer could only be provisional. We are all displaced persons: even the luckiest of us are in some sense asylum seekers, refugees on a journey from somewhere to nowhere, creatures with a sense of a lost Eden, convinced that there must be a haven, a heaven, a place where we belong.

ACKNOWLEDGEMENTS

This is not a memoir, it is a description of the world seen by one pair of eyes. This is not intentionally a science book, although it draws on generations of scientific research. Because it is a book about the nature of address, there is an obvious debt to everybody who shared those places in which I have lived, starting with the family into which I was born, and of course the family that I have now. But an address book that begins with a small room and ends with a universe incurs a much wider indebtedness, one which extends to the books I have read, starting with the Arthur Mee *Children's Encyclopaedia*, and to the people who talked so patiently to me during my decades as a newspaper journalist in two hemispheres, or who unknowingly addressed me through international journals such as *Science*, *Nature* and *Scientific American*. We acquire a world view imperceptibly, from many sources, and it would be preposterous to attempt to identify them all. I have included occasional attributions in the text – the errors however are of my making. Books happen because more than one person believes in them. So I must thank my publisher Nicholas Pearson for his forbearance; my editor Robert Lacey for his patience; my agent Will Francis for making this book happen; and my wife Maureen for putting up with me while I wrote it.